PREFACE

The authors long have maintained that, logically, a printed text in the hands of the pupils is as essential to the proper presentation of the subject of Mechanical Drawing as like texts are necessary to the right teaching of other branches of mathematics in the high schools.

The results of their own experiments in this direction have convinced them of the great helpfulness of such a text to the pupil and to the teacher alike, and of the superiority of this over the oral method of instruction, so long and so generally followed.

Furthermore, it is their belief that, as in other subjects, even approximate uniformity in the teaching of this subject can be secured only by the general use of a uniform text as the basis of instruction.

Technical institutions for years have used such courses, but these are too advanced and too highly technical for secondary school use.

Hence, since no complete courses written especially for the academic high schools are available, the authors, a committee created for the purpose, have endeavored to prepare one, hoping that, by its introduction, the work in this subject may be generally improved and that something like uniformity in it may be secured.

In brief, the course consists of several groups of problems, one or more groups for each school year.

The theoretic and the practical problems of each group have been carefully selected, arranged progressively, and presented simply and clearly in language largely conversational in style, the intention being to talk the subject to the pupil on paper, as it were. In fact, in so far as space would permit, the text was written for self-instruction. For obvious reasons, however, the teacher is not eliminated altogether.

Each group of problems trated text explanatory of such problems, and of the ｌ ciples involved in their solu necessary, in the lower ye problem is accompanied by ing out, and by specific sug

Principles and their prac emphasized all through, fo solving of a stated number essentials to be taught. W the teacher can readily shap the course, so as to add va meet the differing abilities lems, as herein given, he ne detail, but to the methods forth, he should.

Numerous instances pr educational worth of the tr this subject as a branch of guided by instructions syst plying with the exactions pupil will acquire a skill in of practical drafting which cial value to him, should account.

Upon the text of the fir collaborated, while each preparation of that one o which bears his name.

Percy
Berthe
Carl I
Albert
Fred

Chicago, Ill.,
September 30th, 1909.

LIST OF NECESSARY DRAWING MATERIALS.

Set of drawing instruments in a case, comprising:
One pair of 5″ compasses with attachments, namely: A pencil leg, a compass pen, and an extension bar. One pair of 4½″ dividers (a combined compass and dividers will not do). One straight line pen. One 3½″ spring-bow pen (desirable, but not absolutely necessary).

Drawing board, 20″ by 24″.

Triangle, 9″—30 and 60 degrees.

Triangle, 7⅛″—45 degrees.

Tee square, 24″ blade.

Architect's, triangular, boxwood scale, 12″ long.

Paper for first year class, 19″x24″ (making two plates).

Lettering paper for exercises in lettering.

Bottle of Higgin's waterproof black India ink.

Bottle of Higgin's red India ink (optional with teacher).

4II lead pencil, and stick of 6II lead for compasses.

One-half dozen thumb tacks.

Pencil and ink eraser, chamois skin, and sandpaper.

Portfolio for plates, 15″x22″.

Additional Materials Needed for Second, Third, and Fourth Year Work.

Tinting outfit, comprising:
A camel's hair water-color brush, double ended, large and medium size.

Tinting slab or nest of three 2″ saucers.

A stick of India ink or a pan of water-color for tint (sepia or gray).

Two small, handleless, enameled cups for water.

Paper for this work: Whatman's cold ⌐
by 30″, making two plates, each 15″ b⌐
Bottle of Higgin's red India ink.

FIRST YEAR.
INTRODUCTORY TEXT.

The study of constructive drawing or drafting aids in developing the reasoning p⌐ other branches of applied mathematics, str⌐ ventive and constructive ability, and tend⌐ in the pupil the love for systematic, preci⌐ work which will always add to the desira services in any capacity.

Furthermore, a knowledge of its prin value to men in almost every occupatio ingly, this course aims to familiarize the the underlying principles of those methods ical representation in general use which necessary to the practical draftsman.

However, it is not alone a knowledge o and "the how" that makes a draftsman, b knowledge of principles and methods c⌐ the essentials of good workmanship—n CURACY, THOROUGHNESS and NEA of which will be demanded of the student a technical school and in the professional him be accurate, thorough, and neat as principle right from the start, and get the

Directions for the Use of the Impler⌐

Leads.—Pencil and compass leads sh⌐ and 6II respectively, and should be sharp⌐ shaped. Always carry the pencil lightly, ⌐ sure upon such hard leads will make a fu⌐ paper which will be difficult to remove.

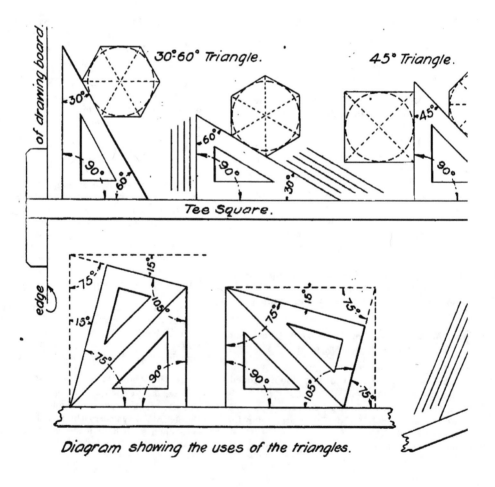

of drawing board.

edge

30°60° Triangle.

45° Triangle.

30°

90°

60°

60°

90°

30°

45°

90°

Tee Square.

75°

15°

105°

15°

75°

90°

75°

15°

75°

90°

105°

75°

Diagram showing the uses of the triangles.

Tee Square and Triangles.—Use the tee square from the left edge of the drawing board always, and in drawing with the tee square and the triangles always use the farther edge, working from left to right or from you.

Use the tee square with the head tight against the left edge of the board for all horizontal lines, and the triangles held against the tee square in the position shown in Fig. 1 for all vertical lines and lines at angles of 15, 30, 45, 60 and 75 degrees to either the vertical or horizontal, as shown in the figure.

By using the tee square held obliquely, in combination with a triangle as a straight edge, parallel lines at any angle may be drawn, and also lines at 15, 30, 45, 60, and 75 degrees to these oblique lines.

By the use of the tee square and the triangles, circles may be subdivided into 2, 3, 4, 6, 8, 12, and 24 equal parts.

Compass and Dividers.—Let the compass and the dividers rest upon the paper with their own weight and manipulate them from the head with the thumb and the first two fingers; do not bear down on them or

grasp the pencil leg or the pen with th[e] while tracing a line. Avoid putting pre dividers in stepping off distances and in lines.

Accuracy.—By accuracy is meant not ness in laying off lines and angles, but al in subdividing them, in connecting point intersections, in drawing perpendiculars, tangents, in locating centers of circles arcs of different radii, etc.

Thoroughness.—By thoroughness is m dent disposition of the pupil or of the make his drawing show clearly and fully information.

Neatness.—Under this heading comes to or detracts from the general appear drawing, such as blots, erasures, uneven shaped letters of varying slants, misspelle Besides this, a symmetrical and agree tioned arrangement of the problems upon necessary in order that, as a whole, th seem to balance, and the effect, from an a point, be pleasing.

Geometric Constructions.

"Geometry is that branch of mathematics which treats of position, form, and magnitude." It deals with the relations of points, surfaces, and solids.

Plane geometry treats of constructions and figures which lie wholly in one plane.

The study and execution of constructions in plane geometry train the pupil to some necessary skill in the use of the drafting instruments, provide him with various constructions of practical value to the draftsman, and give him an introduction to the subject of geometry by familiarizing him with many geometric terms and conceptions. For convenient reference, the definitions of various commonly used terms are here grouped.

Definitions of Geometric Terms.

Lines.—A point is that which has position but no magnitude. It is represented by a dot.

A line is that which has one dimension—namely, length.

A straight or right line is one having the same direction throughout its length.

A curved line or curve changes its direction at each succeeding point.

A horizontal line is one that is level throughout its length.

A vertical line is one that is perfectly erect—i. e., is parallel with a "plumb line."

Parallel lines are those that lie in the same plane and that would never meet if produced. They are equidistant throughout their length.

A perpendicular line is a straight line so meeting another that the two adjacent angles formed are equal. Each of these angles is called a right angle.

A right angle is divided i degrees.

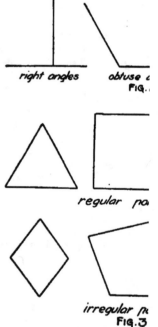

right angles   *obtuse* ⌐
FIG.

regular po

irregular p
FIG.3

Angles.—When any two produced, would meet in a p is called an angle. The two of the angle. The point of n angle is called the vertex.

The size of the angle is the amount of its opening and does not depend upon the length of its sides.

If the opening between its sides is greater than a right angle, the angle is an obtuse angle.

If the opening is less than a right angle, the angle is an acute angle.

Polygons.—A polygon is any angular plane figure bounded by straight lines.

If all its sides and all its angles are equal, it is a regular polygon.

If all its sides and all its angles are not equal, it is an irregular polygon.

Polygons are named according to their number of sides or the number and the kind of their angles; thus, a polygon of four sides is a quadrilateral; a polygon of three sides is a triangle, etc.

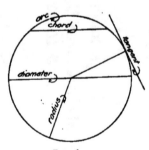

FIG. 4.

Circles.—A circle is a plane figure bounded by a curve, all points of which are equidistant from a point within called the center.

The boundary of a circle is called the circumference.

Any part of the circumference is called

Any straight line having its ends in the ence is called a chord.

Any chord passing through the center eter.

Any straight line from the center to ference is a radius.

Any straight line which touches a circ point is a tangent. It is always perper radius drawn to that point.

Planes.—A plane or plane surface is the straight line connecting any two po wholly within the surface. Example: T a drawing board.

The definitions of the various solids up in connection with the subject of Wc ings.

Lettering.—Ability to letter well and essential to the draftsman as ability to wi those following clerical pursuits.

Legibility is the first requisite of gc This is secured through simplicity and c shape of the letters, uniformity of slant and proper spacing. Practice alone will l

In lettering the plates, print only th the propositions and titles as given. It to do this before the constructions are ways rule guide lines for the locations an of the letters. Letter directly in ink, u nary pen, without first penciling the lett for pretentious headings. Elsewhere th slant lettering, not vertical. Make the tering the same as in writing—i. e., abou

SHOP SKELETON
ITALIC
abcdefghijklmnopqrstuvwxyz
ABCDEFGHIJKLMNOPQRSTUVWXYZ
1234567890

ARCHITECTURAL
VERTICAL OR ITALIC —
abcdefghijklmnopqrstuvwxyz
ABCDEFGHIJKLMNOPORSTUVWXYZ
1234567890

ROMAN
abcdefghijklmnopqrstuvwxyz
ABCDEFGHIJKLMNOPQRSTUVWXYZ
1234567889&

BLOCK
WITH THE TOOLS

Do not crowd the letters or the words. The space between two words should never be less than the width of the capital H of the alphabet used. Do no lettering outside of the border lines.

Examples of Alphabets and Numerals. Fig. 5.

These are standard alphabets and they give the correct shapes of the letters and numerals. **These shapes should be learned.** From these alphabets, other and more decorative ones may be easily derived, but on ordinary drawings, ornate printing is out of place. As previously stated, for general use, inclined letters are preferable because they are easier to make free-hand, hence more rapid of execution, and in them slight differences of slant are less noticeable than would be slight variations from the upright in vertical letters. Use Shop Skeleton for general detail work and, later, the Block, the Roman or the Architectural styles for titles, as is appropriate.

Widths.—Notice that **W** is the widest letter (about as 5+ : 4); that **M** is the next (about as 5— : 4), and that the **J** and the **I** are the narrowest (the former about as 3: 4), and that the others are uniformly of the width of the **H** as a standard. ·

Text of Problems.

Note.—Immediately following the problems here given for Plates 1, 2, 3, and 4 there are twelve others from which substitutions may be made or extra ones given, at the discretion of the teacher.

Plate 1. Fig. 6.

Lettering Exercise.—Before laying out this plate or the three following ones, the pupil is to print care- fully, in class, as a freehand lettering exercise, first,

a Shop Skeleton alphabet of small letters als (Fig. 5), and then the "propositions" o of the problems about to be drawn. This i on a card of the prepared lettering pape ness of shape and uniformity of slant of also suitable spacing of the letters and o are the things which the pupil must tr When this card is done, the plate may according to the following directions:

General Directions.—The sheet of dr is 12″ by 19″. Lay one of these on the dr with its longer edges about horizontal. I square over this, with its head tight aga edge of the board; its blade will then be Bring the upper edge of the sheet to the of the blade of the tee square. Then fast firmly by placing thumb tacks in the corn(in Fig. 6.

With the tee square and a triangle (line ¼″ from the edge on three sides ((right, top, and bottom) to measure from trim to.

From the upper right hand corner of tl set off a point 1¼″ to the left and 1″ fixing a corner of the border line of the pl Fig. 6. From this corner point, draw line 15″ long and a vertical line 9½″ in le the ends of these lines draw the other lin to complete a rectangle 9½″ by 15″.

Divide the longer side of the rectang 5″ divisions and the shorter side into t visions. Then, with the tee square and draw from these points horizontal and v

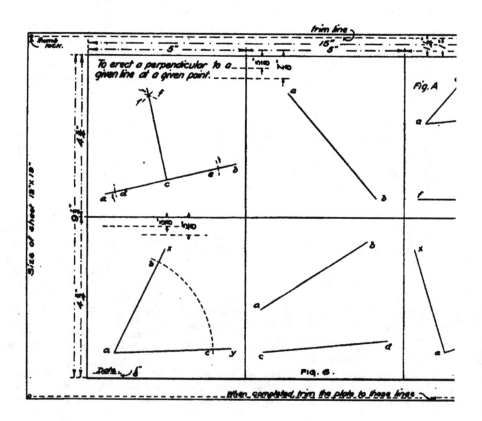

subdividing the large rectangle into six smaller rectangles each 4¾" by 5". Later, in each of these rectangles, we will construct one of the problems.

Next, the pupil should print, at the top of the spaces, the statements of the problems. For these he must rule (in pencil only) the necessary guide lines, as shown in Fig. 6. In this printing he should strive to improve upon the work of his lettering exercise.

In the particulars of size and of position of the problems the pupil must be careful at all times. Drawings that are disproportioned to their surroundings or that are so placed that the plate seems unbalanced, look badly. They should be redrawn. And where there are several problems on the plate, each should be drawn of a size agreeable to the others. These are among the requisites of good workmanship.

All succeeding plates in geometry will be laid out as just described, and all plates of the first year will be of the same size as this one, that is, 9½" by 15" inside of the margin.

In problems where no solution or but a part of one is shown, the pupil must follow carefully the directions given and thus work out the solutions for himself. In doing this, mark each point described, by the letter given to it in the text, otherwise the solutions will be impossible.

The pupil should refer to the definitions in the introductory text for the meanings of unfamiliar terms in the following problems which are not therein explained.

Problem 1.—To erect a perpendicular to a given line at a given point.

Let a-b be the given line and c the given point. With c as the center, using any radius convenient, on the given line, set off d and e equidistant d and then e as centers and with any what greater than d-c, draw arcs intersec f. Draw c-f. This is the perpendicular practice, the draftsman would draw this with the triangles.

Problem 2.—To bisect a given line.

Let a-b be the given line. With ce ends a and b, using a radius greater tha a-b, draw arcs intersecting in points c an site sides of the given line. Connect c an in which the line c-d cuts the given line point of bisection—i. e., the center of the

Problem 3.—To transfer an angle.

Let c-a-b be the given angle, Fig. A. draw a line, as f-i, Fig. B, and assume an k, as the vertex of the required angle. In the vertex, a, as the center, using any rad arc cutting the sides of the given angle in e. In Fig. B, with center k and the radi draw an indefinite arc, h-g. Take d-e, radius, and with g, Fig. B, as the center just drawn, in point s. Now draw a through s. Then s-k-f, Fig. B, will be angle, being equal to the given angle, Fig struction.

Problem 4.—To bisect a given angle.

Let x-a-y be the given angle. With t as the center and with any radius, draw a the sides of the angle in points b and c. points in turn as the centers and any ra than one-half of arc b-c, draw arcs inter side arc b-c in point d. Draw a-d. This the given angle. To continue this proce bisect the angles already formed—use b intermediate point just found on the arc b-c, as the centers from which to draw the necessary intersecting arcs, and then construct the bisectors of these smaller angles in the same manner as that used to find the bisector of the given angle. These bisectors also bisect the chords and the arcs of the angles.

Problem 5.—To bisect an angle the sides of which do not meet.

Let a-b and c-d be the sides of the given angle. By the use of the triangles draw a line parallel with a-b at any convenient distance therefrom, and at an equal distance from c-d draw a line parallel therewith, intersecting the first parallel. Bisect the angle thus formed. Its bisecto also will bisect the angle formed by a-b and c-d.

Note.—To draw lines parallel with a given line by the use of the triangles, or a triangle and the tee square, see the introductory text, Fig. 1. In this case, let a-b be the given line. Place any edge of either triangle (preferably the shorter one) so that it coincides exactly with the given line. Place the other triangle or the tee square against either remaining edge, preferably the longer. Then, holding the second triangle or the tee square securely with the left hand, with the right, slide the first triangle where wished, keeping the edges in contact. Any lines drawn along the edge which in the first position coincided with the given line a-b, will be parallel with that line.

Problem 6.—To trisect a right angle.

Let x-a-y be the given right angle. With any radius and with point a as the center, draw an arc cutting the sides of the angle in points b and c. Using point b as the center and with b-a as the radius, draw an arc cutting the arc b-c in point e. In like· manner,

with point c as the center a an arc cutting arc b-c in po: By these two lines the giv: right angle is the only angle construction.

Inking.—Now, ink the (this, take great care to draw width, as shown in Fig. 7a, dashes uniform in length, as in Fig. 8b. Arcs should inter

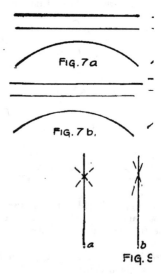

FIG. 7a

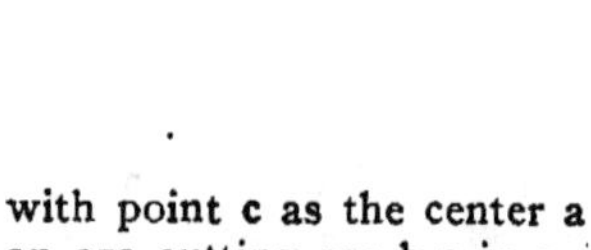

FIG. 7 b.

FIG.

through the point of intersec not as in Fig. 9b, or as in

EXAMPLES OF LINES WITH THEIR USES.

1-For given and result lines in geometric constructions and subdivision I
plates in geometry. For visible edges of objects represented in workii
ings except "shadow-lines". BLACK.

2-For upper and left border lines of plates. Also for "shadow-lines". F
"shadow-lining" Make the lower and the right outlines of surfaces shac
excepting outlines common to two adjacent visible surfaces. BLACK.

3-For right and lower border lines of plates. BLACK.

4-For construction lines in geometry and for all lines representing invis.
of objects. BLACK. For extension lines. RED.

5-For projection lines in working drawings. BLACK.

6-For axes or center lines and for dimension lines. RED.

NOTE:-Arrow heads and figures should always be BLACK, also the sign of feet (') and o

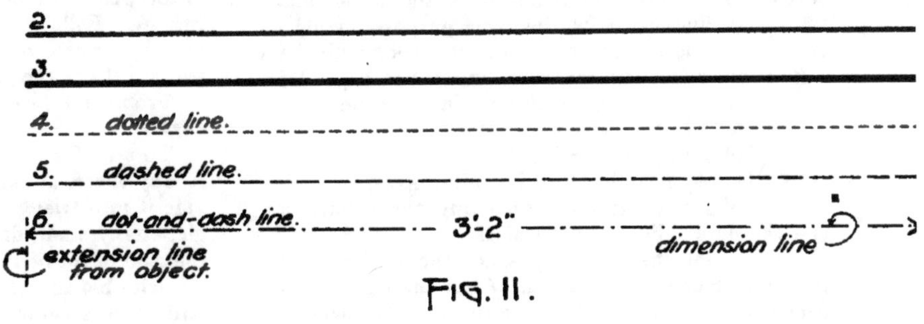

1.

2.

3.

4. dotted line.

5. dashed line.

6. dot-and-dash line. 3'-2" dimension line

extension line from object.

FIG. II.

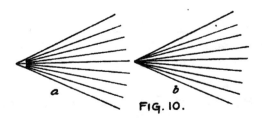

FIG. 10.

angles having a common vertex are best drawn as shown in Fig. 10a, not as in Fig. 10b.

Do not carry the point of the ruling pen close up to the straight edge, and be sure there is no ink on the outside of its blades. Carry the pen vertically, with the set screw away from the straight edge.

In geometrical constructions the lines used in the statement of the problem are the given lines. The lines used in combination with the given lines to determine points or lines are construction lines. The lines obtained as the solution of the problems are the result lines.

Of these, all construction lines should be inked dotted. All given and result lines should be inked solid. All lines used for the same purpose should be uniform in width. Given and result lines should be a trifle heavier than construction lines. See Fig. 11.

In general practice, ink all lines in this order:
1. Dotted arcs and circles.
2. Solid arcs and circles.
3. Dotted straight lines, including axes, etc.
4. Solid straight lines, including the subdivision and the border lines for the plates.

Do not ink the letters by which the various points in the problems were named. After inking, erase all pencil marks and trim the plate to the trim line.

Plate

Do the lettering exercise a lay out the plate in accordanc given for Plate 1. However, division lines as shown in Fig for the problems, then comn

Problem 1-2.—To transfer lels.

In the middle rectangle of irregular polygon of not less Do not copy the figure here

To transfer this polygon, and to construct it in exactly Fig. A, transfer the angle fo sides with either a horizonta as angle 1-b-c, using the me Pl. 1. Then, taking them in thus transferred, 1-b, the rem gon, drawing them parallel sides of the given polygon, b as explained in the note to P: the dividers, make each side in turn, equal in length to the Fig. A. Following this met may be transferred a short and still have the same posit

Problem 2-3.—To transfe angles.

To transfer or reconstruct A, anywhere, at random or in vide it into triangles, as show gle, at any slant, draw the lir the side of b-1 of Fig. A. Ta ter, with b-4 as a radius, draw with l' as a center, and 1-4 a

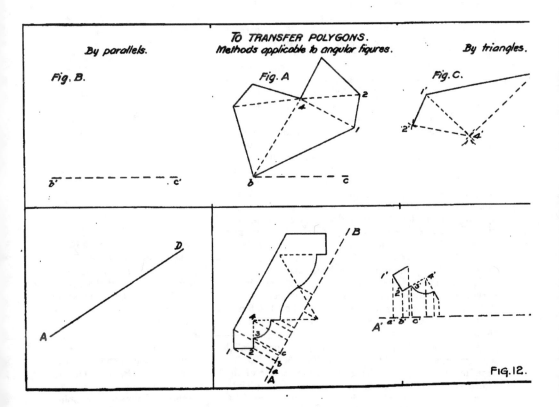

TO TRANSFER POLYGONS.
Methods applicable to angular figures.

By parallels.

By triangles.

Fig. B.

Fig. A

Fig. C.

FIG.12.

cutting the first arc in **4'**. Draw lines **4'-1'** and **4'-b'**, thus constructing triangle **1'-4'-b'**. Now use side **1'-4'** as a base, and, with lines **1-2** and **4-2** of Fig. A as the radii for arcs intersecting in point **2'**, thus construct the triangle **1'-2'-4'**; thereby adding the side **1'-2'** to the figure. By continuing this process—i. e., by building up by triangles—the polygon may be reconstructed. Rule.—The known distances from two points to a third point will determine its location.

Problem 4.—To divide a given line into any required number of equal parts.

Let **A-D** be the given line. Draw, at any angle, an indefinite line from either end of the given line, as from **A.**. On this line set off from point **A** any unit as many times as there are wanted divisions of the given line **A-D**. Connect the last point (**c**) thus laid off, with **D**, the end of the given line. Parallel with the line **c-D**, from all other points on the indefinite line by the use of the triangles, draw lines cutting the given line. The divisions thus made will be equal and of the number required.

Problem 5-6.—To transfer a figure by co-ordinates, that is, by the use of equal measures.

Let the given figure have any outline made up of curves and straight lines. For convenience, but not necessarily, the curves may be circular, and the figure might resemble the cross section of a moulding, as shown, but do not copy this one.

Anywhere without the given figure, about parallel with its principal axis, draw a base line, **A-B**. To this base line draw a perpendicular from each angle of the figure and from the center of each arc. In the sixth rectangle, at a different inclination than that of **A-B**, draw a base line, **A'-B'**, for the new figure. Set off on this base line the distances a'-b', a'-c', etc., always

taken from the initial point, a
a-b, a-c, etc., on the original
pendicular at each of these po
ing each equal respectively to
the like points in the given fig
thus located (1', 2', 3', 4', etc
they are numbered, thus d
figure.

In inking, always keep in
nice lines.

Plate
Lettering exercise; lay ou
problems.

FIG. 13.

Problem 1.—To draw a
given circle and through a
circle, (B) without the circle.

(A) With any radius, dr
assume any point in its circur
point. From the center, c,
length through point a. On t
(a-d), without the circle, e
Construct a perpendicular to
(see Prob. 1, Pl. 1). This is

(B) Assume any point, a
ference. Connect this point
circle, **c**. Bisect this line **e-c**
2, Pl. 1). With **h** as the ce

radius, draw an arc cutting the circumference of the given circle in two points, g and f. These are the points of tangency for right lines passing through point **e**. Draw them exactly, as shown in Fig. 14a, not as in Fig. 14b, or as in Fig. 14c.

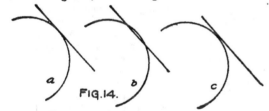

FIG.14.

Problem 2.—To inscribe a circle within and to circumscribe a circle about a given triangle.

Let c-d-e be the given triangle. Let it be irregular.

(A) Bisect any two angles of the triangle (Prob. 4, Pl. 1). The point of intersection of these bisectors (point b) will be the center of the inscribed circle—i. e:, a circle that is tangent to all the sides of the triangle. It is also the center of the triangle. With this center and a radius to any side, draw the circle. The bisector of the third angle should pass through this point. Use this as a test of the accuracy of the first work.

(B) Construct the perpendicular bisectors of any two sides and extend them to their intersection in point a. This will be the center for the circumscribing circle—i. e., one which can be drawn through all cf the angles of the triangle. The perpendicular bisector of the third side should also pass through this point a. Use this as a test of the accuracy of the first bisections.

FIG.15.

Problem 3.—To inscribe a regular pe a given circle.

Draw the given circle, using any ra Draw a horizontal and a vertical or any dicular diameters. Bisect any semi-dia in point c. Using point c as the center, tance to either end of the other diamet an arc from this point (d) cutting the ho eter in a point (e). Then, with point and the distance **d-e** as a radius, draw cutting the nearer part of the circumfere Arc d-f is one-fifth of the circumferen dividers, step off this distance carefully cle, and then connect in order, the poi Extreme care must be taken in each st struction of this and of the following pr light, fine lines; be precise in locating c exact in stepping off the divisions.

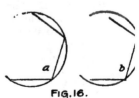

FIG.16.

these polygons must lie precisely in the circumference of the circumscribing circles, as shown in Fig. 16a, not as in Fig. 16b, and their sides must not be "doctored" to make them come out **apparently** correct.

Problem 4.—To inscribe a regular heptagon within a given circle.

Draw the given circle. Draw any diameter, as the vertical one, a-o. With either end of this diameter as the center, trace an arc passing through the center of the circle and cutting its circumference in points b and d. Draw the chord d-b cutting the diameter in the point e. Then, with point d (or b) as the center and with d-e as the radius, draw an arc from point e cutting the circumference in point f. Arc d-f is one-seventh of the circumference of the circle. Therefore, the chord d-f is one side of the required polygon.

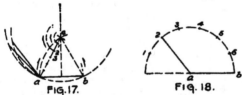

FIG. 17.   FIG. 18.

Problem 5-6.—To construct regular polygons upon given basis.

(A) Let line a-b be the given base (about 1⅛″ long), Fig. 17. Construct an equilateral triangle upon a-b as a base, by drawing arcs intersecting in point c, using a-b as the radius and points a and b as the two centers. Draw and produce the altitude of the triangle indefinitely. Divide a side, as a-c, into six equal parts (Prob. 4, Pl. 2). With point c as the center, trace arcs from the points on side a-c to the altitude produced.

Using, in order, the poi centers and the distances fro extremity of the base (point circles through points a and ly the circumscribing circl having six, seven, eight, etc given length. By this metl six to twelve sides may be here begun are a heptagon

(B) Let line a-b be the gi a-b as the radius and eithe center, draw a semi-circle ar equal parts as there are to b be constructed. Connect the on the semicircle with the ce Construct the perpendicular of the given base, and exte section. With this point, circle through points a and the circumference of this points, thus construct the figure here begun is a hepta

In inking, remember that fine lettering give a drawing

Plate

Lettering exercise; lay o

FIG. 19.

Problem 1.—To draw thr their radii given.

Draw, as the given radii, three lines, **a**, **b**, and **c**, Fig. 19. Draw a line, **1-2**, equal in length to any two of these radii, as line **a** plus line **b**. On this line as a base, construct a triangle, making one side equal in length to line **a** plus line **c**, and the other equal to line **b** plus line **c**. With angle 1 as the center (vertex of the angle formed by the sides **a**), trace a circle with the given radius, **a**. In like manner, with the other angles of the triangle as centers, draw circles, using respectively radius **b** at one center and radius **c** at the other. The circles will be tangent in their points of intersection with the sides of the triangle. The curves must just touch, as shown in Fig. 20a, not as in Fig. 20b, nor miss as in Fig. 20c.

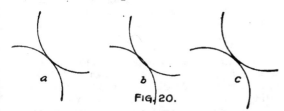

FIG. 20.

Problem 2.—Within an equilateral triangle to draw three circles tangent to each other and to the sides of the triangle. Be extremely careful in the execution of this and of the following problems.

Construct an equilateral triangle (see Prob. 5, Pl. 3). Bisect each angle of the triangle. Extend the bisectors to cut the sides opposite. Bisect either right angle formed by one of these bisectors with the side opposite to the angle which it bisects. The point of intersection (**x**), in which this bisector cuts the bisector of the angle of the given triangle, will be the center for one of the required circles. By means of a

circle, the center of which is the center of **c**, transfer the center, **x**, just found, to the the other angles of the triangle in point These will be the centers for the remain gent circles required, the perpendicular either adjacent side being the proper radi

FIG. 21.

Problem 3.—Within a given circle, to tangent to each other and to the given c:

Draw the given circle, Fig. 21. Divi by diameters into twice as many equal p are number of tangent circles required. diameter. Draw a perpendicular at the adjacent diameter (as at end **b** of diamet intersection (point **a**) with the extend Bisect the angle formed at **a**. The poin this bisector cuts the diameter **c-b**, is the c of the required circles. By means of a c the center **x** to each alternate diameter, the centers for the other required tanger

FIG. 22.

Problem 4.—To construct an approximate ellipse upon axes of given lengths.

Let line **a-b** be the given major or longer axis, and line **c-d** the given minor or shorter axis, the two drawn at right angles and intersecting at their centers, **e,** Fig. 22.

Connect the extremities of the axes, as in line **c-b.** With the center in point **e,** using one-half of the minor axis as the radius, from **c** trace an arc cutting the major axis in point **f.** On the line **c-b** set off a distance from **c** equal to the distance **f-b,** that is, one-half the difference of length between the two axes. Bisect the remainder of this line **c-b,** and extend the perpendicular bisector to cut the major and minor axes (the latter produced, if need be) in points **h** and **j** respectively. From **e,** set off on the major axis the distance **e-i** equal to the distance **e-h,** and on the minor axis, **e-k** equal to **e-j.** From **k** draw indefinite lines through points **h** and **i,** and from **j** an indefinite line through **i.** With point **h** as the center, using **h-b** as the radius, trace an arc from the line **j-h,** through **b** to the line **k-h** produced and so continue, using center **k** for the next arc, and then **i,** and center **j** for the arc

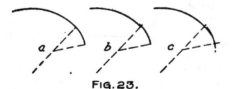

FIG. 23.

completing the figure. The joining of the curves of different radii must be exact, as shown in Fig. 23a, not as in Fig. 23b, or as in Fig. 23c.

FIG. 2.

Problem 5.—To construct axis of a given length.

Let line **a-b** be the given r vide this axis into six equal extend indefinitely its perpen on this bisecting line, up from equal to one-sixth of **a-b,** and (**c-e**) equal to two-sixths of **a**· from each end of the given ax 2 (near **b**), draw indefinite lin **e**. With point **1** as the center, trace, through **b,** an arc from duced. With point **d** as cente line **2-d** produced. With poi curve through **a** to line **2-e** p figure, using point **e** as the cer

Problem 6.—To construct of a circle upon any diameter

If we turn a circle we imi in its appearance. In the d turned it seems diminished. which might be likened to a an ellipse. No part of its out

If, in this turning of the c tion of points in its circum! trace the figure correspondin of the circle. In such a figure

possible to draw is the major axis, and the shortest is
the minor axis. Let us assume that the major and
minor axes are given, lines **a-b** and **c-d** respectively,

FIG. 25

Fig. 25. Upon each, as a diameter, draw a circle.
The larger one we will consider as the unturned circle.
Divide it into twelve equal parts by diameters (see
Prob. 6, Pl. 1). If, now, we revolve the larger circle
about the diameter **a-b** as an axis, the points in its
circumference, as 1, 2, 3, 4, etc., will have the motion
(parallel with the short axis) shown by the arrows
pointing inward. When the points **3** and **8** have come
respectively to points **c** and **d** (the ends of the minor
axis), points **1, 2,** and **4, 5,** etc., will have moved a pro-
portional distance, as determined by lines drawn outward (parallel with the long axis) from the points in
the circumference of the smaller circle, as shown.
Draw a free-hand curve accurately and smoothly
through the points thus found. The figure will be the
required ellipse.

Supplementary Problems.

At the discretion of the teacher, these are to be
used as extra or alternative problems in connection
with those of the four preceding plates, and, like
those, later, in connection with the problems in working drawings and projections.

FIG. 26.   FIG. 27.

Problem 1.—To erect a perpendicular at
of a given line.

Let **a-b** be the given line, Fig. 26. Assum
as **c**, in any convenient position without
With **c** as the center and with the distance
to either end of the line (as c-b) as a radius
arc greater than a semi-circumference cu
given line (point d) and passing through end
d draw a line through **c** and extend it to int
arc just drawn, point **e**. Now, draw a lin
through **e**. This is the required perpendic
practice, the draftsman would draw this wit
angles.

Problem 2.—To bisect a given arc.

Let **a-b** be the given arc, Fig. 27. Make
suitable radius and any length. The soluti
problem is identical with that "To bisect a gi
The bisector of the arc is also the bisector of
and of its angle, as experiment will show. If
it will pass through the center used in tracin
Use this as the test of the accuracy of the v

Problem 3.—To draw through a given po
parallel with a given line.

Let **a-b** be the given line, and **c** the gi
Fig. 28. With **c** as the center, and any c
radius, cut the given line by an indefinite ar
d, as shown. With center d and the same ra
an arc through **c** and cutting the given line i

Taking the distance e-c as a radius, use d as the center and cut the arc first drawn, in point f. Connect c and d, and draw a line through c and f. The lines a-b and c-f are parallel. The angles c-d-e and d-c-f (called "alternate inferior angles") are equal.

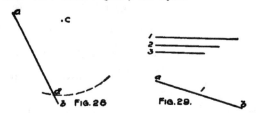

FIG. 28.   FIG. 29.

Problem 4.—To construct a triangle, its sides being given.

Assume three lines, 1, 2 and 3, as the given sides, Fig. 29. Draw, as a base, line a-b equal to 1. With 2 or 3 as the radius, and with end a or b as the center, trace an arc. With the unused given line as the radius and the other end of the base as the center, draw an arc intersecting the one already drawn in point c. This is the vertex of the triangle. Joint point c with the ends of the base and thus complete the figure.

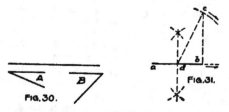

FIG. 30.   FIG. 31.

Problem 5.—To construct a triangle, two of its angles and the included side being given.

Let A and B be the given ably make them acute angles side a line of any convenient l the line 1-2 equal to the giv this construct an angle equal and at the other end construc given angle B. For this, use Prob. 3, Pl. 1. Extend the sic 1-2, indefinitely. These sides pleting the triangle.

Problem 6.—To construct a base of a given length.

Let the line a-b be the giv a perpendicular at either end, to a-b. Bisect a-b in point d. trace an arc from c (the uppe lar), cutting the given base ex a as the center, trace upward as the radius; with b as the drawn in point f, using this s point f as the center and a ra base, a-b, trace an indefinite a below f. With the same radiu as centers, cut this arc in poin ing points a, g, f, h, and b in t ure will be completed.

FIG. 32.

Problem 7.—To construct a base of a given length.

Let a-b be the given base, Fig. 32. With a and then b as the center and with a-b as the radius, draw arcs intersecting in point c. With c as the center and with radius c-a or c-b draw a circle passing through a and b. The base a-b can be set off upon the circumference of this circle six times. As c-b equals a-b, the radius of any circle equals the chord of one-sixth of its circumference.

Problem 8.—To construct a regular octagon upon a base of a given length.

Let line a-b be the given base, Fig. 33. Bisect the base in point c and extend this bisector indefinitely. With c as the center, trace an arc from a or b, cutting the bisector in point d. With d as the center, trace an arc from a, cutting the bisector in point e. With e as the center, pass a circle through the ends of the given base, a-b. The base a-b is the chord of one-eighth of this circle.

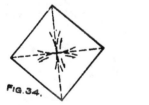

FIG.34.

Problem 9.—To inscribe a regular octagon within a square.

Construct a square and draw its diagonals, Fig. 34. Then, with each angle of the square as the center, and with a semi-diagonal as the radius, trace arcs cutting the sides of the square adjacent to the centers used. Connect the points thus found so as to cut off the cor-

ners of the square. These lines will the octagon. The alternate sides of t coincide with the sides of the square.

Problem 10.—To draw an arc thro points.

Assume any three points, as 1, 2, 3, than in a straight line, Fig. 35. Join 1 These straight lines are chords of tl (see Prob. 2 of this set). Bisect the the bisectors to their intersection in p the center sought. With a radius to ar trace an arc. It will pass through the

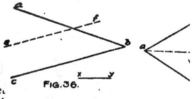

FIG.36.

Problem 11.—To draw within a gi is a given radius, tangent to the sic cle of the angle a-b-c, Fig. 36. A Given angle, en radius of the circle to be as the giv hrough line b-d. The center of t the angle b e te. in this line. A tangent i cle must lie To drawn to the point of tan to the radius given om one side of the an at a distance f e give), a line, e-f, be drav given radius (x as the tting the bisector of that side, and cu n line, will be the center point thus found, center quired. cutti o inscribe an appr

Problem 12.—T within a rhombus.

(A rhombus is a quadrilateral whose sides are equal and whose opposite angles are equal.)

Draw a rhombus, as **a-b-c-d**, of any suitable size, Fig. 37. Bisect each side, as in points **1, 2, 3,** and **4.** Produce these perpendicular bisectors to their intersections with the diagonals of the rhombus, in points **e, f, g,** and **h.** These are the centers required. With a radius from each center to the ends, **1, 2, 3, 4,** of the corresponding bisectors, trace the arcs between these bisectors and tangent to the sides of the rhombus, thus completing the problem.

Definitions of Geometric Terms.

Before entering upon the subject of Working Drawings, we will give some attention to the definitions of the various common geometric solids with which we are about to deal.

Solids.—A solid is that which has three dimensions.

Polyhedrons or plane surface solids: Any solid bounded wholly by planes is a polyhedron. Its edges are the lines of intersection of its bounding planes. Its faces are the plane figures formed by its edges.

A prismatic surface is a surface formed by passing planes through successive pairs of parallel lines.

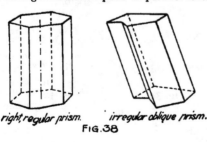

right, regular prism.   *irregular oblique prism.*
FIG. 38

Prisms: A prism is a ʃ closed prismatic surface and bases.

The axis of a prism is tl joining the center of its base centers of circles inscribed ble. If its axis is perpendi right prism. If otherwise, it

If the bases of a right pr it is a regular prism. Any o

right, irregular pyramid
FIG.

Pyramids: A pyramid : a polygon for its base and vertex for its sides.

The axis of a pyramid line from its vertex to the c ter being the center of the c when possible.

If its axis is perpendicul. pyramid. If otherwise, it is

If its base is a regular po mid. Any other pyramid is

Pyramids and prisms ar of their bases, as triangular

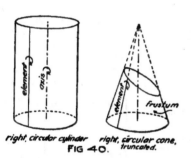

right, circular cylinder   right, circular cone,
FIG 40. *truncated.*

Curved surface solids: Cylinder:—The surface described by the movement of a straight line about a
curve without change of direction (i. e., always remaining parallel with a fixed straight line and passing
through a fixed curve) is a cylindrical surface. A
cylinder is a solid bounded by a closed cylindrical surface and two parallel planes, the bases. The common
kind, the right, circular cylinder, may be considered
as resulting from the revolution of a rectangle about
one of its sides. It resembles a prism, excepting that
its bases are circles.

The side about which the rectangle was revolved
becomes the axis or center line of the cylinder.

Any straight line in the cylindrical surface is an
element of that surface. All elements are parallel
with the axis.

Cone: The surface described by the
a straight line, one end of which is fix(
other follows a curve, is a conical surfac
a solid bounded by a plane and a closec
face. The common variety, the right (
may be considered as resulting from the
a right triangle about one of the sides ac
right angle.

It resembles a pyramid excepting th:
a circle.

The side about which the triangle
becomes the axis or center line of the cor

The vertex of the triangle becomes
the cone.

Any straight line from the vertex to t
element of the conical surface.

Sphere: A sphere is a solid bounded
surface all points of which are equidis
point within called the center. This sph(
may be considered as resulting from the
a circumference about any diameter.

Truncate: The process of cutting off
solid, as of a prism, a cone, etc. See Fig.

Frustum: The part of any solid rer
the top has been cut off, as at a cone, a
It always contains the base of the solid.

Working Drawings.

Working drawings are those conventional representations intended to show clearly and completely the shape, the dimensions (in detail and in general), and the arrangement of parts in an object to be made, accompanied by such explanations and directions as are needed in its making and finish.

Although from an ordinary pictorial representation of an object certain general peculiarities of its form and structure may be readily understood, yet the precise facts of proportion and of construction necessary to its making are lacking. Usually, in such a drawing, its edges and its dimensions are not shown in their true relations, nor its surfaces in their true shapes, while its internal construction may not be evident at all.

If, however, the imagined object be pictured, both visible and invisible parts, upon two or more surfaces assumed in positions parallel respectively with its several dimensions (its length, its breadth, and its thickness), the views thus shown will give the information needed for its manufacture.

Views.—Two such views are always necessary (more are often needed), one taken directly from above, called the "top view," and the second taken directly from in front, called the "front view," the latter of these being always drawn directly below the former. When a side view is used to show more fully the facts of the object, it is drawn directly to the right of or to the left of the front view.

Lay Out.—Fig. 41, working drawing of a hip roof, and chimney.

In the illustration, first, the axes or center lines of the three views were drawn. Then, symmetrically upon these, the top, the front, and the side views were

constructed and in the orde the views were inked and sioned and named, the scale stated, the title or heading p drawn. This is the usual o pupil to follow in his work.

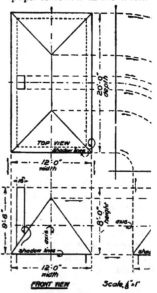

In the figure, notice: Th of the object is very differe

That from all three tog derstanding of the object i two of them, and that no on to make possible the object

That the invisible as well as the visible parts are drawn, the former in dotted lines, and that what is visible in one view may be hidden in some one of the other views.

That the top view shows width (distance from left to right) and depth (distance from front to back).

That the front view shows height and also width.

That the side view shows height and depth.

That each dimension of the object appears in two views of it, and that each view presents two differing dimensions; hence, if these facts of an object are known, by combining them correctly the views of it may be laid out directly and independently of each other. Or if any two views are laid out, the third view may be found by carrying from the views already drawn the unlike dimensions, and combining them.

If, in Fig. 41, the side view is to be found from the two others, the height is carried over from the front view, and the depth is swung about from the top view, and combined with the height, as shown, the center for this revolution being located wherever convenient.

Data.—The facts of proportion and of structure of objects, when obtained or calculated, may be recorded as written description only, as pictorial sketches illustrative of and accompanied by description, and as dimensioned free-hand working sketches with specifications attached.

Invariably, working drawings are made from such notes or records, which are always necessary if the objects are at all complex, as are articles of furniture, machinery, buildings, etc.

In these several ways the problems will be presented in this text in order that thereby the pupil may become familiar with practical methods.

The problems will comprise a "stateme but not always, accompanied by a "spe laying out," and "suggestions for solu needed.

Scales.—Working drawings are const: "full size" or "to scale." In the first cas ing is the actual shape and of the actual d the object or part. In the second case, tual shape is shown, the dimensions a actual, but are at a stated ratio to the ac expressed in "inches per foot," which mea inch or certain fraction of an inch of dim drawing is the equivalent of one foot in or "on the ground," as 1½" equals 1', or etc.

On the implement, the scale, the units (⅛", ¼", ¾", etc.) when large enough to are divided into twelve equal parts in o this means, inches may be laid off on th the proper proportion.

The scale to which a drawing is made be stated.

Here the pupil should begin his work ferring to the following paragraphs on Inl sions, and Titles when directed, as his wc

Inking.—In inking working drawing edges of the objects are made with solid invisible edges are indicated by dotted construction lines drawn from view to vi "dashed." Ink the extension and dim after the views have been inked and th inserted. See Fig. 11.

If, in each view of the object, the r lower outlines (when these are not edge two visible faces) are made heavier th outlines, an effect of shadow is produced which strengthens and much improves the appearance of the drawing. These are called "shadow lines," and the process is termed "shadow lining." See Fig. 41. Do not shadow line developments.

Dimensions.—Generally, the pictures of an object, such as are shown in a working drawing of it, are not, in themselves, sufficient for the making of the object, even though the scale of the drawing is stated. Although these representations show the arrangement and the relative proportions of its parts, etc., they are incomplete and of little practical use as working drawings until, upon them, the precise dimensions of the object, as a whole and in part, are fully marked. The workman must not be left to "scale" or to guess at the dimensions of any important or specially designed parts.

If the fact is kept in mind that every form and its every part have three dimensions, the process of dimensioning will be readily understood. It is evident that somewhere upon the drawing each of these dimensions must be shown at least once. Dimensions should be put only upon distances, linear and angular, which are shown in truth.

The points between which measurements are taken must be clearly indicated. When, as is usually best, the dimensions can be placed wholly outside of the drawing, these points are shown by the use of parallel "extension lines" drawn squarely outward therefrom. Then, perpendicularly between these extension lines, the dimension lines are drawn and the dimensions are inserted, at their centers, if possible. The extension and dimension lines are generally drawn in red ink to distinguish them from the lines of the object, the extension lines being "dashed" and the dimension lines

"dot and dash," or often, to solid, broken only for the dim

To emphasize the initial black arrow-heads are made a sion lines.

Dimension and extension l as to conflict as little as po their importance, the figures perfectly legible and so printe without inverting the drawir ways in which the figures sh horizontal as the lower edge

Titles.—Although on sho dom much time spent in the n and of ornate border lines, yet always be executed tastefully they may be, in order that th pleasing finish.

The title should be place needed to maintain the bala plate, and not, as an afterthou anywhere."

In the design of titles, th compactly grouped in order t a glance. Do not string the Generally, for titles, use verti to slant letters. Do not com

DESIGN
FOR A
DRAWING EASEL
SCALE - 3"=12"
(with tools)

APARTMENT · BUILDING ·
AT · 639 GREENWOOD · AVE ·
FOR · HENRY P · SMITH ·
ARTHUR JONES · ARCHITECT ·
Chicago · July · 1909 ·
(free-hand)

CYLINDER
FOR A
6"×8" VERTICAL ENGINE.
May, 1909.   Lane Tech.
Scale, 1½"ᴎ1'
(free-hand or with tools)

in a title or on the drawing, such as very heavy and fancy letters and light, plain ones. Also, for like parts of a drawing, and for a set of drawings of the same subject and kind, the titles and general lettering should be uniform in style. In many offices, a set form of title is stamped or stenciled upon the drawings to secure uniformity and to save the draftsman's time.

Emphasize the important statements by larger letters or by stronger treatment, and subordinate the less important ones.

Before working out the title upon the drawing, it is often well to sketch it elsewhere, experimentally, to determine the best size and style of letters and the best arrangement for the various rows of words. When these rows of words are centered upon a common vertical axis, as shown in Fig. 43, begin with the middle letter of each row and work both ways. Of course, such designing is best done in pencil first and inked later.

Do not try to make showy or elaborate border lines having more interest than the drawing itself, perhaps; keep them simple and appropriate. Fig. 43 gives a few suggestions for the design of titles, of border lines and of corners. Only occasionally are these latter called for.

Plate 5.

This plate will be given wholly to a careful duplicating of Fig. 11, "Examples of Lines and Their Uses," the printing to be executed free-hand.

This is to be done on a sheet of the regular drawing paper, the plate to be of the same size as the previous ones, 9½″x15″.

The exercise comprises the title, the statement of uses, and the examples of lines.

The alphabet used shall slant or vertical. The neces rows of words shall be 5/16 the rows of words shall be The examples of lines shall b cal axis of the mass of prin that of the plate.

The height of the small the height of the capitals sha erly thereto; see Fig. 5.

A slightly ornate border li wished.

Work for correct shape a ters, and for good spacing Do one sheet in pencil, as doing a second directly in ini not first write in pencil and ing with a pen.

To facilitate his work as must acquire the ready use of

Text of Pro

Of the seven following pl to secure his year's credit in lettering exercises called for and any two of the others tha For the experience to be gai them all.

Plate 6.

Problem 1.—A rectangular prism with development.

Make a working drawing of the rectangular prism shown in Fig. 44.

Evidently, each view of this prism will be a rectangle, and yet it is also evident that no two of these

views (such as the front and the top or the side views) have one center line or —i. e., in common. By thus centering t common axis, their like points will fall into the right relation, as shown by tl struction lines in Fig. 41; those marked Therefore, in this and later problems

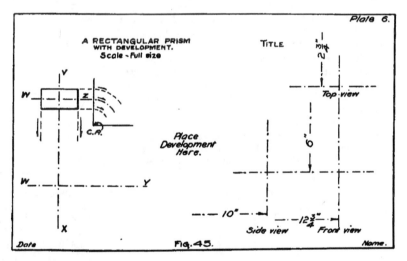

views will be alike. Hence, to show all its appearances, the top, the front, and the side of the prism should be drawn. As the prism is small, these views may be drawn the full size of the faces. The object is assumed as upright with its wider face to the front.

Now, how should the views be arranged? Note, in Fig. 41, that, as before stated, each two adjacent

center lines first, and then construct the objects upon them.

Location.—Fig. 45. Where conver case, 1¾" to the right (R) of the left the plate and parallel therewith, draw Assume this as the common axis for th front views. Unless otherwise directed, or plan should always be drawn first, when this is possible. The top view of the prism is a rectangle ¾″ by 1½″, and, in accordance with the assumed position of the prism, the longer sides of this rectangle should be horizontal. With its center line of width, **W-Z, 3¼″** down (D) from the upper border line, let us construct this rectangular top view, drawing one-half of it on each side of the vertical axis.

From a point 6½″ D, measured on the vertical axis, draw a horizontal line, W-Y. This will be the common axis for the front and the side views.

With the triangle and tee square, carry down the width of the prism from the top view, as indicated. Then set off one-half of the height of the prism (1¼″) above and below this horizontal center line just drawn, and so construct the front view.

Again referring to Fig. 41, note how, by revolution, the depth in the top view is carried about and down, and combined with the height carried across from the front view, to work out the side view. In

like manner, construct the si suming the center for the r and 4¼″ D.

In order to show the rel ners of the prism in all view always giving the same na views, as illustrated in Fig. sion of points, the pupil sho do this with all problems.

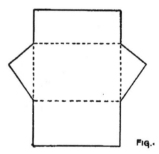

Fig.·

Development.—Frequentl an object is desired, requiri faces be laid out, giving w ment" or unfolding. In th should be arranged in accor positions in the original ob next to each other in the ob the development. Fig. 47 s a right triangular prism an pyramid.

In the space to the right after completing the second velopment of the rectangula

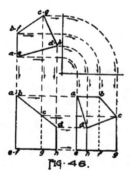

Fig. 46.

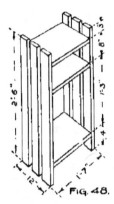

FIG. 48.

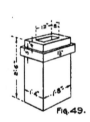

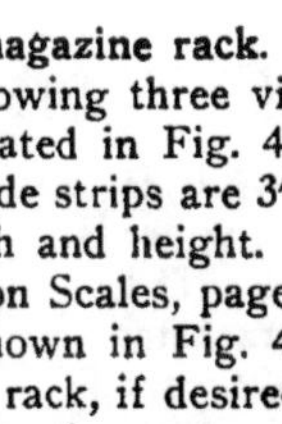

FIG. 49.

Problem 2.—A rectangular magazine rack.

Make a working drawing, showing three views of the simple magazine rack illustrated in Fig. 48. All parts are 1″ in thickness. The side strips are 3″ wide. The front view is to show width and height. Scale, 1½″ equals 12″; see paragraph on Scales, page 26...

Note.—The brick chimney shown in Fig. 49 may be substituted for this magazine rack, if desired.

Location.—Proceed in exactly the same way as that followed in working o. t the previous problem.

The axes for the several views are to be located where shown in the location chart, Fig. 45. Construct the side view directly from the dimensions given, and not by revolution, as was done in Prob. 1.

Before inking, read the paragraphs on this subject, page 26. Note that, in the side view of Prob. 2, the shelves are, in part, invisible; hence such parts should be indicated by dotted lines. Also, using dashed lines, ink the construction lines extending

from points in one view to like points views. Design a neat title for each draw the scale in this.

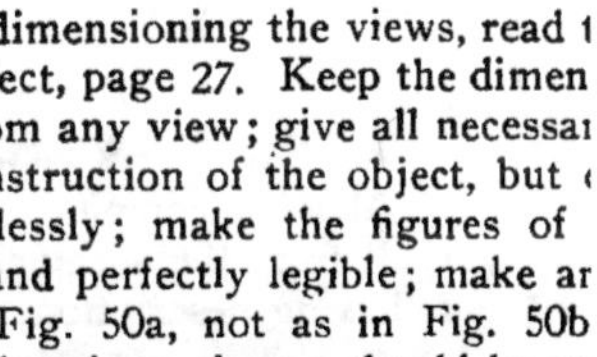

FIG. 50.

Before dimensioning the views, read 1 on the subject, page 27. Keep the dimen or more from any view; give all necessar for the construction of the object, but them needlessly; make the figures of good size and perfectly legible; make ar shown in Fig. 50a, not as in Fig. 50b that your drawings always should be so complete in their information that they derstood perfectly by the workman in objects represented.

Lettering Exercises.—Too much em; be put upon the fact that skill in free-ha1 a necessity to the draftsman. This h(only by much and careful practice. 1 working drawings, aside from the title sions, require but little lettering. Henc exercise of fifty words, to be done on lettering paper at other times than the ods, will be required with this and with succeeding plates.

Besides the alphabets and the num lections to be printed are the importan paragraphs, the definitions, etc., from the introductory text, at the direction of the teacher.

The practice should be upon one alphabet only, until that is mastered. Of these, the Shop Skeleton is the most generally serviceable.

Plate 7.

Problem 1.—A rectangular shelf.

Make a complete working drawing for a simple shelf from the accompanying sketch, Fig. 51, and the following description:

The shelf is 2′ long, 9″ wide and 1″ thick. It is attached to a back or wall piece 2″ below the upper edge of the latter. This is 2′ by 1′ and 1″ thick. The shelf is supported by two right triangular brackets, each 8″ by 8″ by 1″. These brackets are set 1½″ from the end of the shelf. The shelf is horizontal, with the brackets visible in the front view. Scale, 1″ to 6″. Use for this, the scale 1″ to 1′, doubling the readings.

Location.—To begin with, as in the previous problems, draw the axes where specified. In the side view, the axes of width and of height intersect 6½″ R. and 6⅝″ D. In the front view, the axes of length and of height intersect 2½″ R. and 6⅝″ D. From these two views, find the top view, assuming the center of revolution to be 5″ R. and 4¼″ D.

First, as shown in Fig. 5: where required. From this the levels of the various par ting off the length of the sl the position of the brackets, in the location fixed by the :

Now, from the side view, width of the shelf, the outer Then, from the front view, c shelf and the back and th of the brackets, thus workii the top view, the brackets a be inked accordingly.

Note.—The brackets and ornamented in a simple way is encouraged thus to add to his work.

Problem 2.—A rectangul

Note.—A knife box, a c object of equal difficulty, m problem here given, at the It is suggested that the pup for some one of these article a working drawing therefro

Make a complete worki tangular stool from the giv sketch, Fig. 53.

Notice that, in the figure, only one-half of the front view is shown. The pupil is to draw the whole stool. Scale, 1″ to 6′. The stool is horizontal, with both supports visible in the front view.

Location.—The axes intersect in the top view, 12¾″ R. and 3″ D.; in the front view, 12¾″ R. and 6⅞″ D. The center of revolution for the end view is 10⅜″ R. and 4½″ D., the revolution to be to the left. Draw the axes of arrangement, then the top, the front, and finally the side view.

In inking, shadow-line and dimension the views as before, showing all invisible parts by dotted lines.

Plate 8.

Problem 1.—A flight of steps from a model.

(Left one-half of the plate.)

The pupil is to make, in shop perspective, a dimensioned, free-hand sketch of a flight of three or more steps from the school room model. See note below.

This sketch is to be made as preparation for the class work at other time than the drawing period. From this sketch a working drawing of the object, showing three views, is to be made in class.

A suitable scale, the proper location of the axes and of the views, is to be determined by trial, by the pupil. The sketch is to be drawn in the upper left hand corner of the plate. It is to be inked and dimensioned.

For most of the plates, in order to assure good placing of the work, and also to save the pupil's time, the locations of the problems are specified. In these the positions of the views are fixed by locating their axes. In like manner, the pupil should fix the positions of the views in problems for whic[ ] tions are given—i. e., by drawing the views first.

It is not well for the pupil to beco[ ] upon "full specifications" for laying out upon "full explanations of solutions." ever particulars of this kind are lacki[ ] pected to rely somewhat upon his own k[ ] judgment as to "what to do" and "how [ ]

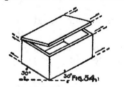

Note.—In shop perspective the retr[ ] edges of the object are drawn parallel proaching, as they retreat, as is the cas[ ] spective. Also, in rectangular objects the retreating edges are often drawn lengths instead of fore-shortened—i. e., minished because turned away. Freque[ ] termed "isometric sketches," isometric equal measures." Such sketches are ea[ ] using the 30 degree triangle as shown in tice the difference between Fig. 54 and [ ] drawn first in shop perspective and the[ ] spective.

Problem 2.—A waste box or similar[ ] object.

The pupil is to prepare a sketch as in[ ] problem, and to make a complete workin[ ] the object therefrom, also showing the [ ] fore.

Plate 9.

Problem 1.—A right circular cylinder from the pupil's sketch.

The pupil is to make a working sketch from the object and a working drawing from the sketch. A working sketch is simply a working drawing made free-hand, and to no definite scale.

Notice that, in this problem, the side view is unnecessary, as it is the same as the front view and so presents no additional facts. Make the cylinder full size, diameter 1¾", length 2½", axis vertical.

Location.—The top view is a circle. Place its center 1¾" R. and 3½" D. In the front view the lower base is 8" D.

In the space between this problem and Prob. 2, after the latter is done, lay out the development of the cylinder. Allowance for this development must be made in placing the second problem. By experiment with a piece of paper the pupil will discover that the development of a right cylinder is a rectangle. The width of this rectangle is the height of the cylinder. To obtain the length, set off, as a unit, the chord of one-twelfth of the base of the cylinder, twelve times, taking this chord from the top view. See Pl. 1, Prob. 6. Complete the rectangle and attach the bases.

Problem 2.—A caster.

Note.—The cylindrical wall pedestal shown in Fig. 59, or the cast iron bracket shown in Figs. 60, 61, may be substituted for this, if desired. Descriptions below.

The Caster.—Make a complete working drawing of the caster, from the side view, Fig. 56, and the description. Scale, full size. The sketch, Fig. 57, gives

the thickness of the fork, and of the tread or cylindrical sur

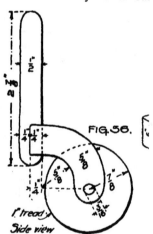

FIG. 56.

f' tread
Side view

Location.—Draw the side the others. Place the center 7⅜" D. and 10⅝" R.; the rolle in the figure. The location pupil is to determine.

The drawing is to be dir lined, as usual. To shadow-l diameters at 45 degrees, Fig. the shadow-line arc down to t the circle a distance equal to the shadow-line. With the sa circle, using the center (c'), d other diameter, below, and to outside cylindrical surface, and above and to the left, to show an inside cylindrical surface, as shown in the figure.

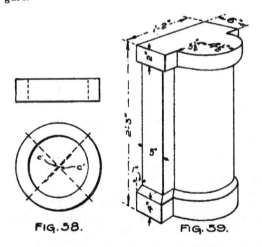

FIG. 58.   FIG. 59.

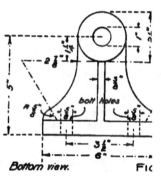

Bottom view.   FIG

The Wall Bracket.—The facts of t given in a **bottom view** and in a sketch view shows the appearance of the ob at from beneath. Draw this view, pla outline 8″ D. and the vertical axis of the

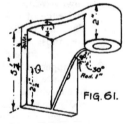

FIG. 61.

The Pedestal.—The top and base of the pedestal each project 1″ beyond the shaft at the front and at the sides, but not at the back. The cylindrical parts are less than one-half of a cylinder. The center and the radius for the cylindrical portions must be found from the dimensions given, by the method given in Supplementary Prob. 10. The whole base is 5″ high; the face 4″, the slope of its upper edge being at 45 degrees. All other dimensions are given in the sketch, Fig. 59. The front is to show in the front view. Locations of the views to be determined by the pupil. Scale, 1½″ to 12″.

The front view will be drawn abov the same way that such a view is draw view. From this bottom view and the formation given in the sketch, complete side views, the latter being placed dire of the front view, as is usual. Scale, on

Plate 10.

Problem 1.—A regular pentagonal pyramid.

Make a working drawing of a right, vertical, regular pentagonal pyramid having an altitude of 3¼″ and a base inscribed in a circle of 1⅛″ radius. See Prob. 3, Pl. 3.

The pyramid is to be so placed that an angle of the base is at the back, and so that, in the front view, three of its lateral faces, or sides, are visible, two equally each side of the third, the middle one.

Draw the top, the front, and the side view, and the development or pattern of the pyramid. Scale, full size.

Location.—The vertex in the top view is 1¾″ R. and 2½″ D. In the front view, the base is 8¼″ D. The center of revolution for the side view is 3¼″ R. and 3¾″ D. The vertex in the development is 6¼ R. and 2¼″ D. From this point draw a vertical line downward, making it equal in length to the rear, lateral edge of the pyramid, as shown in the side view. This line is the true length of all the lateral edges— i. e., those from the vertex to the base. It will be the left side of the development. Using it as a radius and its upper end, the vertex, as center, draw an arc of indefinite length. On this arc set off the length of one side of the base (taken from the top view) five times, and join the consecutive points. Also connect them with the center used. Then attach the base. This is the method used in laying out the development of all regular pyramids.

Problem 2.—A street lamp. Fig. 62.

Note.—The garden flower-box shown in Fig. 63 may be substituted for this object. Also a pyramidal waste basket, a lamp shade, or a tile chimney top might be used in its stead.

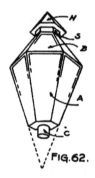

FIG. 62.

The Street Lamp.—The la of two regular hexagonal pyr base, the longest diagonal of v tude of the inverted pyramid upright pyramid (**B**) is 9″. 1 cut off (truncated) 1′ 5″ fro parallel therewith. In like truncated 6½″ from its base. 3″ in diameter, projects 3½″ portion of (**S**) supporting the The altitude of the hood is 3′ its base is 7″. Draw three vi 12″.

Location.—Three faces of visible in the front view, two middle one. Draw the top vie mids first. This will be a reg plementary Prob. 7. Place th 2½″ D. and 12¾″ R. The p position of the other views. These will represent the late

mids. Next, draw the front view of the two large
pyramids complete, and truncate them where and as
directed. Add parts **S** and **C** in this view and indicate these parts in the top view. Now draw the top
view of the hood and then its front view. Finally,
from these two views of the lamp, obtain the side
view. In this latter, only two faces of each pyramid
will appear and these will be equal.

The whole front view and one-half c
are given, as shown in the accompa:
plate, Fig. 64. The top view is to be
the side view is to be found. In the f
axes intersect, 2½" R. and 6½" D. Tl
the prism is inclined at 60 degrees left t
as shown. The center of revolution foi
is 4¼" R. and 4" D.

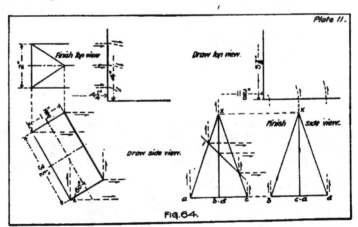

FIG. 64.

Plate 11.

**Problem 1.—A right, isosceles, triangular prism
in an oblique position.**

In all previous problems we have drawn only
simple views of the given objects. In this problem
we are to work out the views of an object that is
tipped—i. e., is in an oblique position.

Begin with the front view. Letter
and also the corresponding points evi
complete top view. The ends of the p
and, inasmuch as the top view of the
shown directly above the front view
must follow that the top view of the lo
a like figure to this, drawn directly a
view of the lower end. Then, to find tl

a vertical line up from each point in the front view of the lower end, and, in the top view, extend each corresponding incomplete lateral edge of the prism to its intersection with the proper vertical line, as determined by the lettering. Connect the points thus found.

Side View.—If now a line from each point in the top view be carried about and down, and a line be carried across from each corresponding point in the front view, the point in which these lines of the same letter intersect will be the point of like letter in the side view. In this way all points may be found. By connecting them, as in the other views, the side view of the prism can be constructed.

Problem 2.—A square pyramid, truncated.

The front and side views of the whole pyramid are given, also the position of the cutting plane in the front view. The problem is, to construct the top view and to show in it, and in the side view, the shape of the surface, called sectional surface, made by truncating the pyramid—i. e., cutting the top from it, as shown.

The diagonal of the base of the pyramid is 2½" long and its altitude is 3¾". The pyramid stands corner ways—i. e., with one angle to the front, and so that the two adjacent faces are visible equally. The base, then, in each of these views, will show the full length of the diagonal—namely, 2½". Also in each of these views, the vertical axis of the view will coincide with the intermediate lateral edge.

Reversing the procedure followed in Prob. 1, in finding the top view, carry lines up and about from the points in the side view to their intersections with lines of the same letters carried up from the front view. Thus the top view can be constructed.

In like manner, the po: edges are cut by the obliqu from the front to the top, a example, in the front view point 1, the top view of thi view of the same edge x-a, ε view; also the side view of : view of the edge of x-a, an front view of the point. A be found and the sectional

Plate

Problem 1.—A regular, !

Make a complete worki: assuming it to be 27" long the thickness of its sides 2".

Location.—The tile is visible in the front view, t of the third, the middle fa: the top view, 2¼" R. and : the front view 2¼" R. an: revolution for the side view

Note that but two faces :

Fig. 65.

Problem 2.—A regular !

Make a working drawin; the tabourette according to Fig. 65, and the following d

The height under the top is 20″. The width of the faces is 7½″. The top is 1″ thick, and overhangs the faces 1½″. The side openings are 2½″ by 1′ 5″, but may be changed to suit the taste of the pupil. The sides are 1″ thick. The position of the object is the same as that of the tile. Scale, 1½″ to 1′.

Location.—The axes intersect in the top view 12½″ R. and 3⅜″ D. In the front view t D. The center of revolution for the side R. and 4⅞″ D. Swing to the left. Fir in the usual manner.

Note.—The side openings may be in a simple way and the top may be d some geometric figure, if desired.

SECOND YEAR.

ORTHOGRAPHIC PROJECTION.

Introductory Text.

All those drawings which the pupil made in his first year, representing geometric solids, various articles of furniture, etc., and which we called "working drawings," were, in reality, simple orthographic projections of the objects drawn; orthographic projection being the correct mathematical term for the manner in which they were represented.

Hence, we are not about to take up a wholly new and strange subject, but rather we are to continue our study of a subject which is already somewhat familiar to us, but under another name. We will look at our drawings henceforth, as problems in mathematics instead of merely as pictures, but the process of their making will remain unchanged from that used in the previous year.

All drawings in which are shown the facts of dimension and of arrangement of parts in objects of a structural character, are termed "constructive" or "working drawings." See the paragraphs in "Introductory Text of the First Year," headed "Working Drawings."

The means utilized to obtain the views necessary to show these facts of such an object, is called orthographic, right, or true projection.

The subject is applied descriptive geometry, which treats of the graphical solution of all problems involving three dimensions—i. e., by picturing their relation as determined by reference to certain planes and lines having fixed positions. Its principles are studied abstractly, and problems requiring their practical appli-

cation are given in order correctly, to make and to re

Definition.—Orthographi of representing objects by two or more intersecting a planes, by perpendiculars points of the objects.

Planes.—In accordance practice, these planes, calle are assumed as above (the fore (the vertical plane, **V**) is helpful, at one side or er vertical plane, **SV**). The assumed to the left or to th convenient or the more usei the co-ordinate planes of glass box, or show case, often used to illustrate th the top of the box represer the side, **SV**; (Fig. 1). F fourth view of the object n of an oblique plane of proje

Views.—The image of is called the top view or pl view or elevation; while tl side view or side elevation lique plane is called an aux

Lines.—The perpendicul object to the several plane: images are obtained, are cal jectors. The line of inter with **SV** is called the hor (**HA**), while that of **V** wit axis of projection (**VA**). with **V** is called the auxilia

Sketch showing the relative positions of the object
and of the planes of projection, and the process
of obtaining the working views or projections of
it on them.

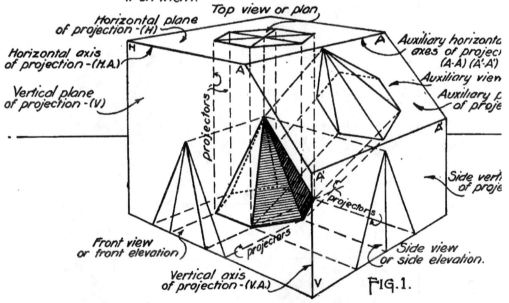

FIG.1.

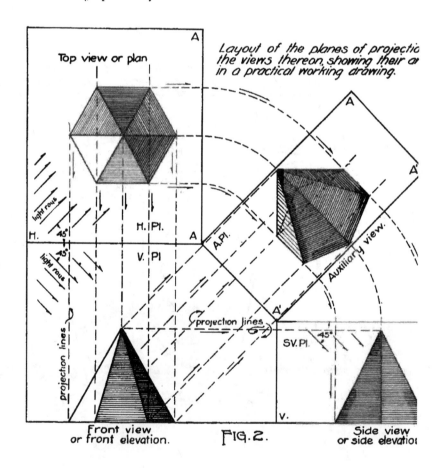

Layout of the planes of projectio
the views thereon, showing their ar
in a practical working drawing.

Top view or plan

A.Pl.

H. Pl.

V. Pl

light rous

45°

45°

light rous

H.

A

A

Auxiliary view.

A'

projection lines

projection lines

SV. Pl.

45°

V.

Front view
or front elevation.

FIG. 2.

Side view
or side elevatio

Layout.—Omitting the oblique plane of projection, imagine the other planes with the views thereon, revolved upon their axes of projection until they lie in the plane of the drawing board, Fig. 2a. Then all above **HA** will be the horizontal plane (**H**); all immediately below **HA** will be the vertical plane (**V**); and all to either side of **V** will be the side-vertical plane (**SV**). Figure 2 shows the layout of the combination of planes shown in Fig. 1.

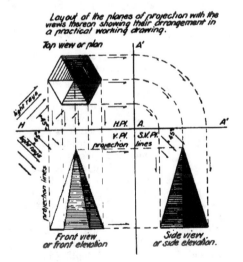

Layout of the planes of projection with the views thereon showing their arrangement in a practical working drawing.

FIG. 2a

Furthermore, in the first instance, Fig. 2a, the front view (on **V**) will be found to lie, point for point, exactly beneath the top view (on **H**), while the side or end view (on **SV**) will be situated right or to the left of the front view.

Also, the projectors from like poin views will appear as continuous lines to the intervening axis of projection.

Reading.—As stated in an earlier pa ing to read working drawings is of eq with learning to make them, and to the is even more so. Most business and pr are called upon to read such drawings themselves not draftsmen, they must eral, what drawings of this kind should understand what, in particular, a giver tell.

This ability the pupil will acquire re make it a practice to master each prob ing to the next. By this means he will understanding of the relation of the va each other—i. e., what facts of the ob shows.

To secure full knowledge of the f tion, position, location and dimension from a working drawing of it, it is pla formation shown in all views of it mu gether, Fig. 3.

Its form and construction are under general shapes of the various views a the different possible appearances of th

Its position is determined by notin gular relations of its axes, bases, and ed of projection, **HA** and **VA**.

Its location is established by tak distances of its axes, bases, and edges fr projection.

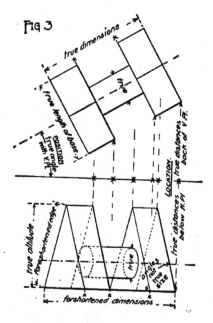

Fig 3

Its dimensions are learned from those of its axes, edges, and faces which are parallel with some one of the planes of projection. It is only upon such that dimensions may properly be placed.

When the teacher has gone through the introductory text with the class to this point, it will be well for him to take up with them, at the blackboard, various simple exercises in the determining of the positions and locations of points, lines, and planes, etc., illustrative further of the subject of Reading, and also

of the following Rules. The
to the text explanatory of th
and start the class on Plate 1.

The remaining paragraph
Text should be referred to as
occasion requires, as directed
text of the problems.

Rules of Pro

1. Any two adjacent proj
always lie in the same projec

2. If two points are actua
jections must always be conn

3. If two or more lines or
lel, their projections must al
one plane of projection.

4. If two lines or edges in
must always intersect in the p
of actual intersection.

5. The projections of a
plane to which it is perpendi

6. The projections of a p
which it is perpendicular will

7. All positions and loca
distances—are alike determine
jection.

8. All vertical distances—
shown upon **V** and **SV**. All
e., widths and depths—must
width (distance from left to r
on **V**. And depth (distance f
be shown on **SV**.

9. In any two views one
mon, the others differing. As
and front views; depth is co
views; height is common to

10. All distances from, and angles with, the axes of projection, shown upon one plane, are distances from and angles with either opposite plane of projection.

11. All movements parallel with one plane of projection will appear in truth upon the planes with which they are parallel. Upon the opposite plane they will be indicated by straight lines parallel with the intervening axis of projection.

12. True lengths of lines and edges, also true sizes of angles and planes, are evident only when they are parallel with a plane of projection, and then only upon that plane with which they are parallel.

13. Points and edges must be "followed up" and located by name, but like points, in all views of an object, must be given the same sign—i. e., either letter or number.

Inking.—All visible edges should be represented by solid lines.

The edges nearest the various planes usually will be visible in the view on the plane to which they are nearest.

All invisible edges should be represented by dotted lines. They should be equal in width to lines used to represent the various visible edges.

The edges farthest from the various planes, usually will be invisible in the view on the plane from which they are farthest.

Edges beneath (usually farthest from **H**, as determined from the front and the side views), will be invisible in the top view.

Rear edges (those farthest from **V**, as determined from the top and side views), usually will be invisible in the front view.

Edges farthest from **SV** (as detern top and the front views), usually will the side view.

All edges connecting visible and i or two invisible points, are invisible.

All projection lines should be mad and in black, and drawn only where a principle is shown.

All construction lines and planes s dot-and-dash lines in black.

All axes of projection, when shov made solid, in black.

All axes of form, center lines, and lines should be made dot-and-dash, ir tension lines should be drawn dotted, i

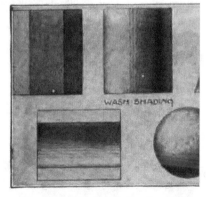

WASH SHADING

Tinting.—Wash shading is a means out the relative positions of the visib various objects drawn, as an aid to th derstanding of their form. It is the most expeditious and least trying on the eyes of all methods.

The light is usually assumed as coming from above, behind, and from the left of the draftsman, the projections of all rays forming angles of 45 degrees (opening to the right) with the axes of projection in all planes. Then, the surfaces will be light or dark in proportion to the directness or indirectness of the light upon them, Fig. 2.

All plane surfaces, as those of prisms, pyramids, bases of cylinders, etc., are tinted flatly—i. e., uniform in tone throughout. Developments should be tinted flatly and of the same tone as the lightest surface of the original object.

All curved surfaces, as those of the cylinder, cone, and sphere, are graded from the line or point of high light. In such forms allowance should be made for atmospheric reflection, because of which the depth of shade is lessened on that part of the surface diametrically opposite to the high light, Fig. 4.

All views of the same surface in a given position must have the same tone. A change of its position may change its tone.

Washes of stick India ink, of gray or of sepia water color may be used with equally pleasing results.

Working drawings of furniture, apparatus, etc., are not to be tinted, merely shadow lined.

RIGHT LINE

Line Shading.—This is a different tones to the surfaces ally useful in suggesting the conical, and spherical surface ings intended for illustrative patent office bulletins, and te

The lines are elements of graduation of tones is secur ing and the thickness of the sumed to have the same direc See Fig. 5.

Evolutions.

Simple views of an object are those in which the axes are severally parallel with the planes of projection.

Oblique views of an object are those in which one or more of the axes are inclined to one or more of the planes of projection. Oblique views are the more pictorial, and frequently aid greatly in the understanding of the form and structural peculiarities of an object. Also, in many combinations and in complex objects, s'mple and oblique views of different parts occur in the same figure unavoidably.

The purpose of problems in Evolutions or movements of solids, etc., is to familiarize the pupil with the processes of working out such oblique views, for the practical reasons just stated.

Oblique views may be obtained in two ways.

First Method.—Fig. 6. By transpo view, and from it and the opposite uni deriving the unknown view corres transposed one, and then continuing t the views thus derived. The solutior lems depends directly upon the app truth that all motions made and all d with any plane of projection can be s upon the plane with which they are p and 12. For example: Fig. 6. If th moved to the right or to the left a the movement being parallel with **V**, ing top view will be found exactly th to the right or to the left of the previou of course, exactly above this new fron point, as established by projection lir like manner, from a transposed top sponding front view may be found.

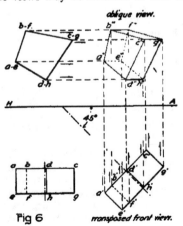

Fig 6

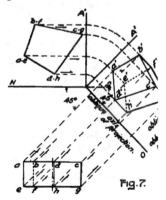

Fig.7.

Second Method.—Fig. 7. By projecting directly from the simple views of an object to an oblique or auxiliary plane. This plane must be perpendicular to one of the planes of projection, usually the vertical, but may be inclined at any angle to the other. Its intersection with **V** and **H** establish the auxiliary axes of projection. The oblique view is then found just as a simple view on **SV** would be obtained. See also Figs. 1 and 2.

Note.—In either case, follow out the method and work out the views: do not try to imagine their appearance.

Evolutions of Lines.—Oblique views of an edge or other line do not show its true length. Simple views do. (Rule 11). If, then, the true length of an oblique edge is wanted, as is often the case, it can be obtained by revolving one of its views until it is parallel with the opposite plane of projection, and deriving the corresponding new view of the edge upon that plane by the method explained above. The process of finding the true length of an edge and of a plane from their oblique views will be explained in detail later where first needed in the solution of a problem.

Size of Plates.—The size of all second, third, and fourth year plates is 12"x17" inside of the border lines, unless otherwise specified. Upper and lower margins to be 1¼", and right margin 1½", when the plate is finished.

Lettering.—All plates in evolutions should have the title, "Evolutions of Solids," and those in sections, "Sections of Solids". If, in either subject, developments appear upon the plate, the words, "and Developments", should be added to the title.

Also, throughout this yea fifty words must accompany just as was the practice of t

Plate

Problem 1.—Make a dia ual arrangement of the plan on these projections of an ol Also name all parts of the 2a.

Use for this problem th ramid the top view of whic place of the triangular pyra 8. The altitude of the pyra

Fig. 9.

Location.—In this and i lems, when used in the spec from the upper border line if so noted), and **R** means border line, excepting when of an angle.

Also, unless otherwise st the axes of the objects, as th ifications, are angles with t Thus, 60 degrees L. to H, means, in the first instance

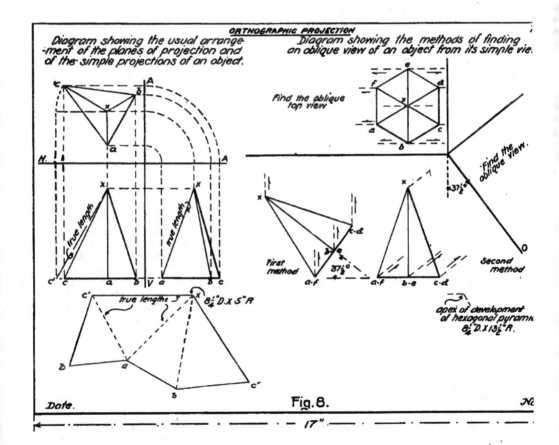

ORTHOGRAPHIC PROJECTION

Diagram showing the usual arrange-
-ment of the planes of projection and
of the simple projections of an object.

Diagram showing the methods of finding
an oblique view of an object from its simple vie

Find the oblique
top view

Find the
oblique view.

37½°

First
method

37½°

Second
method

true lengths   8¼" D. x 5" R

apex of development
of hexagonal pyram
8¼" D. x 13½" R.

Date.   Fig. 8.   N.

— 17" —

angle of 60 degrees with the horizontal plane and opens to the left, while in the second, the axis is at an angle of 45 degrees with the vertical plane, opening to the right, in each case determined from **HA.** See Rules 7, 10, and 11.

Place the horizontal axis of projection (**HA**) 4¼" D. and the vertical axis of projection (**VA**) 3½" R.

In the top view, the center of the circle circumscribing the base of the pyramid, is 2⅜" D. and 2" R. In the front view, the base of the pyramid is 7¾" D.

For the development, if not apparent (Rule 12), the true lengths of the lateral edges (those from the vertex to the base) must be found. This is merely a problem in the evolutions of lines, as mentioned in the preceding paragraphs.

In Fig. 8, Prob. 1, notice that the edge **x-a** is parallel with **SV**, and that its true length is evident on **SV**. This is also the true length of edge **x-b**, the pyramid being isosceles.

Now, if the other edge, **x-c**, be revolved, in the top view, as indicated, about point **x** as a center, until it is parallel with **HA**, hence with **V**, and its lower end in its new position, **c'**, be connected with poin t **x**, which is stationary, this line, **x-c'**, will be the true length of the edge **x-c**. By a like process, the true length of any oblique edges can be found upon any of the planes.

The development of the pyramid is then laid out by the triangulation method of reconstructing a polygon given in Prob. 1-3, Pl. 2, First Year, the edges of the pyramid being the sides of the triangles.

Problem 2.—Make a diagram illustrating the methods of finding an oblique view of an object from its simple views.

Draw what is given as t Fig. 8, and work out the obli lique side view called for. N: ing, when completed, as sho

Notice in Fig. 6 and Fig. found by the two different This is simply because, in b of projection upon which t thrown, have exactly the san of the objects.

It follows, then, in this pr the pyramid in the transpos is inclined at an angle of 37½ tal plane, the oblique or auxi od, must be inclined at 37½ pyramid in its given positio oblique views shall be identi

Location.—Place **HA** 4" I section of the axes of projec the base of the regular hex long, and the altitude of the top view, the center of the b R. In the front view the b Two sides of the base are pe

Draw the simple front an draw the transposed front from it and the simple top v top view. Next, work out th ond method. Finally, lay the space below, with the ver true length of a lateral edge but as the pyramid is regula Ink, print in full, as directe as explained in the paragraf

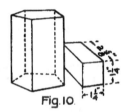

Fig. 10.

Plate 2.—Evolutions of Solids.

Read the whole problem through carefully, first. Then draw the complete statement of the problem before attempting its solution. To start the problems right, make it a practice always to do this.

Given a right, regular, pentagonal prism; its base inscribed in a circle of 1¼″ radius; its axis vertical; its height 3″; its right face perpendicular to **V** and to H.

To the right of this lies a right, square, horizontal prism of the size and in the position shown in Fig. 10, its faces parallel in pairs with the planes of projection. Draw **HA** 6⅛″ D.

(1) For the top view of the pentagonal prism, place the center of the circumscribing circle 2⅛″ R. and 2¾″ above **HA**. Draw the view; a regular pentagon, with its right side perpendicular to **HA**.

Place the point of intersection of the axes of height and of width in the front view of this prism 2⅛″ R. and 2¾″ below **HA**. Draw the view.

Place the point of intersection of the axes of the horizontal prism, in the top view, 4⅝″ R. and 2¾″ above **HA**. Draw the top view; a rectangle, 2⅝″ by 1¼″, having its shorter side parallel with **HA**. In the front view, the axes, intersect 4⅝″ R. and 3⅝″ below **HA**. This view will be a square 1¼″ on each

side. Letter all points in both views o

(2) Redraw the front view of 1 prism with its axes intersecting 9⅞″ same level as in (1).

Redraw or transpose the front view onal prism with its right lower edges 8⅜ below **HA**, the object so tipped that its against the upper left edge of the ho: Letter completely, just as lettered befc

To find the top view corresponding posed front view, use the first method lique views, already explained, Fig. 6.

Inasmuch as the movement of the ol directly to the right, parallel with **V**, this motion by lines parallel with **HA** right from all points of the top view 11). Then draw a projection line up f in the front view of (2), to its interse line of like letter from (1). By this m may be located in their new positions. necting them properly, i. e., as they ar nected (Rule 2), this view may be co

(3) Transpose the whole top view j so that its right and left axis intersect: at an angle of 52½ degrees R. with F left angle of the lower base of the pe in this axis 1⅝″ from **HA**, measured o transposing views of this kind, use 1 method of transferring polygons, Prob 2, First Year. Letter all points, as in ι

Now, inasmuch as this last movem jects has been directly to the right wit level, hence parallel with H, on **V**, sh by lines parallel with **HA** drawn to all points of the front view of (2). (R draw a projection line down from each point in the top view of (3) to its intersection with the line of the same letter (2). By connecting the points thus found, as in the previous views, this view can be completed.

Before inking the figures, their accuracy should be tested in accordance with Rule 3. The projections of all edges which, by construction or arrangement, are parallel, must always be parallel.

In inking, show all projection or construction lines, axes, and invisible edges. Omit the letters placed on the figures, but print the title, then tint.

The teacher may substitute a regular heptagonal prism for this problem at his discretion.

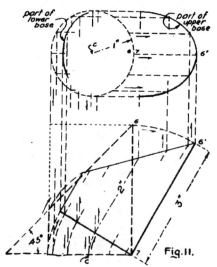

Fig. 11.

Plate 3.—Evoluti

Problem 1.—Read throug first.

Given a right, circular cy titude 3"; axis cut 2" abo inclined at 45 degrees R t 4¼" D.

(1) The cylinder is to its axis is inclined 60 degre with V. First draw the top object, assuming it to be ve shown in Fig. 11.

In the temporary front v of the base, is 1¾" R. and 3 the top view, 1¾" R. and view, a circle, into twelve these points to the base in th them.

Now, in the front view, t 7, to the required angle, ca found, to this new front viev

The top view correspond view is found by the First lique views (Fig. 6), an is previous plate excepting, in t establishing the outline of th freehand curves, forming straight lines, forming angul

(2) Transpose the top its axis makes an apparent a V. and that point C is 6¼" Use the coordinate method f all points in the outlines of th out the corresponding view.

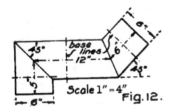

Scale 1″=4″ Fig. 12.

Problem 2.—Cylindrical pipe.

Given the front view of a 6″ cylindrical pipe composed of three sections, having the dimensions and arrangement shown, Fig. 12.

(1) Draw the front view with the lower opening centered 11½″ R. and 3½″ below **HA.** Draw the axes of the sections first.

(2) To work out the top view, draw the circle which is the projection of the vertical section on H, and divide this into twelve equal parts. From these points carry elements (See "Definitions" page 24) the whole length of the pipe. Then the top view of the oblique ends of the second and third sections will be found in the same way in which the ellipses representing the bases were obtained in the previous problem.

(3) Lay out the developments of the sections in order, the lowest to the left of the plate, centered both ways, on a base line 2¼″ above the lower border line and with the axis of each section vertical. It is well to draw the development of the middle section first in the center of the space, and then to place the development of an end on each side.

Suggestion.—Note, that in the front view, as the axis of the pipe is parallel with **V,** the true lengths of the several sections are shown. Hence, in this view

all elements must be evident in their t
they are parallel with the axis of leng
tion, and therefore are parallel with **V**
laying out the developments, spacing
(a twelfth of a circumference apart),
extremities correctly located in referei
line, as determined by the right section
the sections where shown. This is ar
the coordinate method of transferring
Prob. 5-6, Pl. 2, First Year.

Developments.—The practical value
often unknown and unthought of by
manner of sheet metal working of tins
paper box making begins with the lay
developments of the things to be m
ducts of the first include hoods, flues,
grain chutes, bay coverings, gutters an
cornices and architectural ornament in
in copper. Work of the second are ar
coal hods, dust pans, scoops, funnels, p
and all varieties of tinware; and the thi
conceivable shape of box made from car
are but a few of the practical uses of
Lastly, the making of developments is
exercises for increasing the pupil's :
working drawings.

In inking this plate omit all proje
struction lines excepting those used to
lique position of the cylinder in the
Before tinting, practice, using Fig. 5

Plate 4.—Furniture Drawi

Some piece of furniture such as a l:
for chemistry, physics, biology, etc., a
bookkeeping desk, or teachers desk, or

board, should be selected or assigned which is similar to that shown in Fig. 13.

The pupil must make a preliminary free hand constructive sketch of the model chosen, inserting all necessary dimensions, and then make his working drawing from the sketch, laying it out to such scale as will best suit the size of the plate.

This working drawing sh
facts, important details of
mensions needed for the ma
specifications as to the kind
and the nature of the finish
line; do not tint the drawing

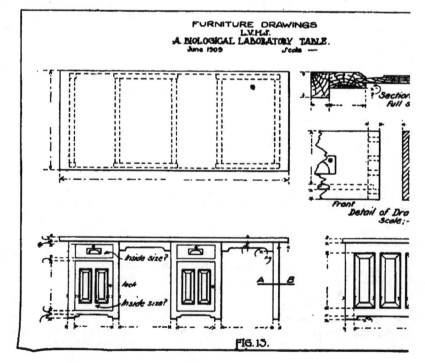

FIG. 13.

Sections.

The views of the exterior of an object seldom make clear its internal structure, or its concealed arrangement and operation of parts. In such a case it is customary to remove some portion of it. This is done by passing an imaginary cutting plane through it in such a way as will be most helpful, and then representing what is thus exposed. Such drawings are called sectional views or simply sections. Fig. 14, is a partial section of a common faucet; note the interior mechanism, of which no outside view could give any idea.

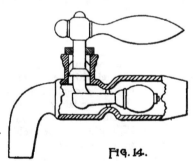

Fig. 14.

Also, in irregular and oblique forms, sections give directly, truths of dimension, angles, etc., not otherwise evident nor more readily found, and the use of cutting planes simplifies the solution of both abstract and applied problems in penetrations, all of which will be shown later.

Sectional views are very common, as illustrations, in all technical and scientific books, and they are invaluable in all structural work.

Such cuttings are usually made eith coinciding with the long axis or length or perpendicular, or oblique thereto, bein respectively, longitudinal sections, cross tions, and oblique sections.

Also, in structures the principal axes o fixed positions as in machinery, vesse etc., vertical and horizontal sectional vie solute necessity to their construction. in the case of a building, these latter, t sectional views, taken at different levels, eral floor plans, without which the struct be erected.

In finding the true sizes of the secti in the following problems, use the secor determining oblique views, namely, by tl lique or auxiliary planes parallel with planes.

Definitions.

Traces.—The point in which a line plane is the trace of that line in that pla

The line in which a plane intersects ɛ is the trace of the first plane in the secc

Plate 5.

Problem 1.—(1) Given an irregular, angular pyramid; base horizontal; angl the circumference of a circle of 1⅛" tude 3".

In the plan (top view), the vertex 1⅞" D. It must not coincide with the circumscribing circle. In the front el front view), the base is 8" D. Draw t then the front and side elevations. One b-1, Fig. 15, is parallel with **V**. The front elevation of another, **b-3**, are perpendicular to **HA**. The third edge, **b-2**, must be oblique to all planes. Letter all points.

(2) Pass a plane perpendicular to **V**, and inclined 45 degrees R. to H, cutting the axis, **b-c**, 1¼″ above the base. Point c, the center of the base, is the point of intersection of the bisector of its angles. (See the definition of "the axis of a pyramid"; also see Prob. 2, Pl. 3, First Year).

The sectional or cutting plane, being perpendicular to **V.**, its projection thereon (its front view) will be a line (Rule 6) inclined downwards to the right at an angle of 45 degrees to **HA**.

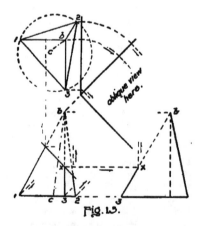

Fig. 15.

(3) To find, in the plan, the resulting sectional surface, or the traces of the plane in the faces of the pyramid, project from the points in which the lateral

edges are cut in the front ele corresponding edges. In a si points to the edges in the Pl. 10, First Year.

But point **x**, in the front cannot be transferred to the because the edge and the ve: it coincide. If, however, the front view to the side view up, it can be revolved to it: view of this edge without di

(4) Now, by connecting plan and in the side elevatio tional surface can be drawn.

(5) Find the projection nitions, first year text) of t lique or auxiliary plane ass parallel with the sectional su by the "second method" of as fully explained and illustr 'Evolutions", Fig. 7, and in Pl. 1, Fig. 8. The problem tical with this one.

Locate the center of rev axes of projection where the views already drawn. The view of the whole pyramid. points of the sectional surfa their proper edges in the o them. This view of the se its true size. Why?

(6) Develop the whole tional line, or trace of the c

attach the sectional surface and the base. This development is to be drawn in the space below the problem, the placing being left to the pupil to determine.

For its development, the true length of every edge in the object must be known; Rule 12. The sides of the base are in true length in the top view. The true length of the lateral edge a-1 appears on V, and that of edge b-3, on SV. The true length of edge b-2 alone must be found. Revolve or transpose this edge to a position parallel with V, using the vertex in the top view as the center, and establish the corresponding front view on V, just as has been explained in the text accompanying Plate 1.

To determine the correct position of the section points, note or find their positions on the true lengths of the edges, and transfer these distances to the edges of the same letters in the development. When these points are connected in the development, the resulting sectional lines should equal the lines of the same lettering in the auxiliary view of the sectional surface. Check the work in this way before inking. Attach the sectional surface and the base.

In inking, in all views and in development, dot that portion of the pyramid which has been removed, as well as the invisible edges of the frustum.

Note.—Any other form of pyramid may be substituted for this, at the discretion of the teacher.

Problem 2.—Given a right, circular cylinder; diameter 2½″; altitude 3″; axis, vertical. The axis, in the plan is 11⅞″ R., and 2⅜″ D. The lower base, in the elevation, is 8″ D.

In the front elevation, a plane perpendicular to V and inclined 60 degrees L. to H, cuts the right element 5/16″ from the top. Also, on the same side, an-

other plane, perpendicular to V, and inc grees R. to H., is passed, cutting the ri 5/16″ from the bottom.

(1) Draw the plan, the front, and th tion. To begin with, divide the circumfe circle representing the plan into twelve and project lines from these points throu and side elevations, these lines being eler cylindrical surface. See, Definitions; Fir

(2) Find the sectional surfaces in th tion, these not being apparent in the the do this, carry, from the front elevation, t which the elements are cut by the section this view, to the elements of the same l side view, and then connect them. The be elliptical curves, to be drawn freehand.

(3) Show the true size of these "se auxiliary planes to the left and to the front elevation. Refer to the preious for information, if needed.

(4) Develop the cylinder, and locate traces of the cutting planes. See Prob. 1 Year, and also Pl. 3, Second Year.

Draw as many of the twelve element velopment as were cut by the planes, to them, from the elements of the same l front and side elevatoins, their section p

Note.—A cone may be substituted fc der, if desired.

Plate 6.

Sections and Evolutions of Solids, w: ment.

Problem 1.—Given a cube having an ε one diagonal of the top of which is perp and the other parallel with V.

Pass a plane perpendicular to **V** through all the faces, cutting each face in a line parallel with one diagonal and midway between this diagonal and a corner of the face. As shown in Fig. 16, in the front view, the trace of this plane passes from the center of the upper edge of the right face downwards to the center of the lower edge of the left face, intersecting the intermediate vertical edge at its center. The section will be a regular hexagon, although not apparent as such in any of the following views.

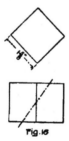

Fig. 16

The pupil must determine the proper spacing for all views and developments on this plate, using one-half the plate for each problem. As stated in the first year text, it is not well for him to become dependent upon "full specifications" for "laying out" his work, nor upon "full explanations of solution." Hence, whenever particulars of this kind are lacking, he is expected to rely upon his own knowledge and judgment as to "what to do", and "how to do it". In this way alone will he come to a working command of the subject. Explanations will be given when a new method or a new principle is involved in the solution

of a problem. These the pupi remember together with those

(1) Draw the top and fron letter all corners in each. Th pass the plane as described ab in which it cuts the edges, and top view.

(2) Tip the front view sc is inclined 60 degrees **L.** to 1 sponding top view, which, of above the new front view and the right of the first top view.

(3) Turn the top view j right and left axis is inclined and work out the correspondi explanations are desired, refer Pl. 2.

(4) Develop the object ir problem. The true lengths of Develop the whole cube first, line and attach the sectional

Problem 2.—Given a regul: side of base, 1¼″; two sides of to **V**; altitude 3″; axis of py: and inclined 60 degrees **L.** tc with **SV** cuts the axis 1¼″ ab

(1) To construct the fron position specified, the top and object placed vertically migh tipped, as in Prob. 1, Pl. 3, bu preferable.

Draw the axis at the requir lower end (c) draw a perpendi 17, draw an arc somewhat grea ference, using a radius of 1¼″, the side of the base of the pyramid. As but two lateral faces of the pyramid will show in the front view, from point **3** on the arc, (opposite **c**), with the radius just used step off **3-2**, **2-1**, and **3-4**, **4-5**. Connect these; the figure will be an auxiliary view of one-half of the base upon a plane

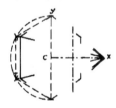

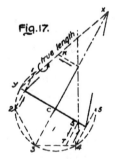

Fig.17.

parallel with the base in its required position; hence its true size. From it the width of the base is measured in the front view. By a like process to that used to obtain the plan in Prob. 1, Pl. 3, in combination with the front view, the top view is found as shown in the

figure. Construct the views, and lette them. Then, from these two views, (view, lettering its points also.

(2) Now, pass the cutting plane. allel with **SV**, hence perpendicular t **H**, its plan and front elevation will be The true shape and the true size of t show in the side elevation; Rule 12.

(3) Develop the frustum; first, h oping the entire pyramid. The true le: al edge is nowhere shown. This ca: revolution, as previously done, or b: The latter method is the more servic use will demonstrate.

In the front view, **x-c**, coinciding w the true length of the altitude. In the a semi-diagonal, is the true distance fr the axis to all the angles of the base, **x-c-y** is a right angle. These constitu a right triangle, th third side cf whi: nuse, is the lateral edge, **x-y**. Theref combined, as shown in the front view tremities are connected, the connectir the true length of all the lateral edge regular pyramid. Use it in laying ou ment, as shown in Plate 1, Prob. 2.

To this measuring line, **x-y**, the sec be carried, as point **k**, and their true f several edges thus can be determined. them to the edges of the same letter : ment.

The points in which this sectional sides of the base of the pyramid, are lo ing this section line, **s-t**, in the auxilia from which the front view was obtai

Attach the sectional surface and the base. Check the section lines in the development with the sides of the section shown in the side view, before inking. They should agree exactly.

Plate 7.—Sections.—Practical Problems.

The pupils knowledge of the subjects of Evolutions and of Sections will have little value until it is turned to practical use in the solution of problems such as the draftsman meets daily.

Figures 18 to 26, inclusive, are dimensioned sketches representing various common objects in the world of mechanics. The pupil is to be assigned the problem of making, on this plate, working drawings of at least two of them, as particularly described below.

view the cutting plane to ⟨
the crank pin and of the cra
In inking, shadow line the
"section line" the section
but not that through the pir
23 are sectioned lined to 1t
cast iron. Steel is shown
wide spacing of the section
steel.

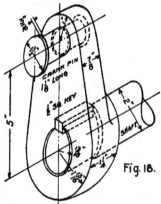

Fig. 18.

Figure 18. An engine crank. Draw the front view, the top view, and a vertical section, as a side

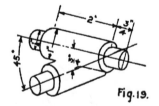

Fig. 19.

Figure 19. Skew pins.
skew pins shown in Fig. 1
of pins are horizontal and
bisector of the angle betw⟨
L to V.

Figure 20. Cylinder he
part in section and part ful
front and the top views co

section. The head is fastened to the cylinder by six ½″ stud bolts uniformly spaced. The gland cover and seat for it are approximate ellipses having a 4″ major and a 2¾″ minor axis. Two ⅜″ stud bolts hold it in place. Indicate the screw threads in the manner shown. Scale, ¾ full size.

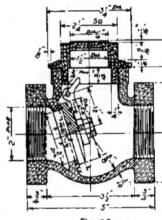

Fig. 22.

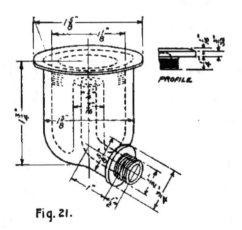

Fig. 21.

Figure 21. Oil cup for an engine. Draw the top view with the cover removed and the stem to the right. Draw the side view in full, and as the front view, a vertical section coinciding with the axis and parallel with V. Draw the cover separately. The threads of the cover are 16 per inch and of the stem, 13 per inch. Indicate and mark them but do not try to draw them. The "profile" shows the shape of the cover in section at its edge. Scale, double size.

Figure 22. A flap valve. Given th section of a flap valve, to draw the fron views entire. Also draw three views separately, the cap, the valve seat cover, The valve operates as follows. The wa left to right and raises the valve cove supply is cut off the cover drops of it thus preventing the water from flowin

Valve proper. Each end of the val hexagon. Thus a w. .nch can be ap valve screwed to a pipe. The section is the long diagonal of the hexagonal en is divided into two compartments by a which is a 2″ hole. The threaded end an inside diameter of 2⅜″, this being t ameter of a 2″ pipe. It has 11½ thread

Notice that the valve seat has a raised boss slightly larger than the valve cover. This boss is ground to insure a close fit of the cover.

Valve seat cover. This is circular, 2¼″ in diameter, flat on the lower side and convex on the upper. The small projection in the center is ⅜″ in diameter and has 16 threads per inch. This projection passes through the valve lever and is secured to it by a hexagonal nut having a long diagonal of 13/16″, and a height of ⅜″.

Valve lever. This is ½″ wide, has an enlarged circular end, 1″ in diameter, and is held in place by a 3/16″ pin.

Valve cap. This is square on top. The circular part is 2¼″ in diameter and has 16 threads per inch. The inside of the cap it cored out to reduce its weight. Scale, full size.

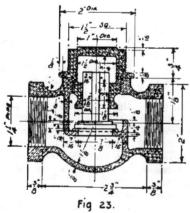

Fig 23.

Figure 23. A check val‹ the assembled valve and alsc Its operation is similar to th

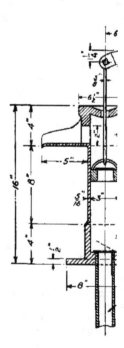

water flows from left to ri plunger. When the flow cea prevents the return of the ν

The valve consists of three parts, the body of the valve, the cap, and the valve plunger. The inside diameter of the ends should be 1⅝″, as that is the outside diameter of a 1¼″ pipe. The number of threads at the ends is 11½ per inch. The longitudinal section shown is taken through the long diagonal of the hexagonal end. A cross section of the valve plunger above the valve disc is a cross. Scale, full size.

Figure 24. A lift pump. Draw three views of the pump including the section shown. Also draw three views of the valve and rod, and of the handle. The inside width of the spout is 2″. Scale, 2″ to 12″.

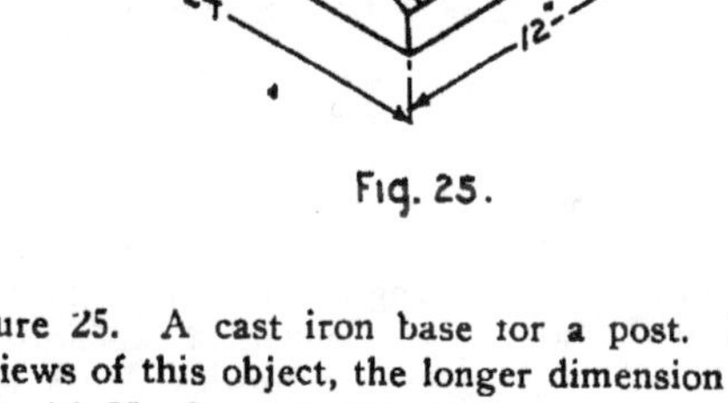

Figure 26. A cast iron cap for a post. views of this object with a vertical sec the center, perpendicular to V, as the side with the longer dimension parallel witl 1½″ to 12″.

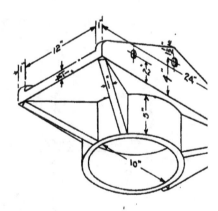

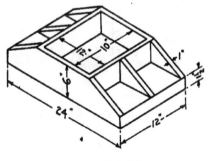

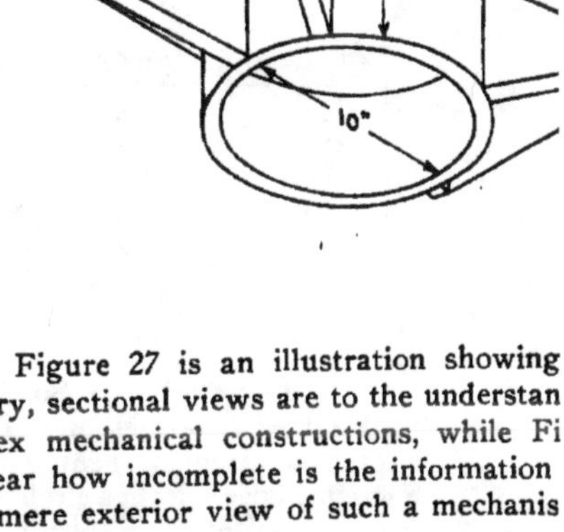

Fig. 25.

Figure 25. A cast iron base for a post. Draw three views of this object, the longer dimension to be parallel with V. Scale, 1½″ to 12″.

Figure 27 is an illustration showing sary, sectional views are to the understan plex mechanical constructions, while Fi clear how incomplete is the information a mere exterior view of such a mechanis

The teacher may add similar probl above selections at his discretion.

Cylinder for Steam Pump.

Sectional View.

Exterior View

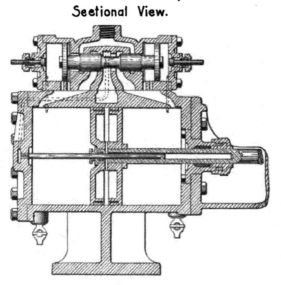

Fig. 27.

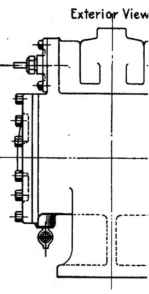

Fig. 28.

Plate 8.—Sections of Solids with Developments.

Problem 1.—Given an oblique pyramid; base horizontal, an irregular pentagon, with one "re-entrant" angle, included within a circle of 1¼" radius, altitude 3". For the meaning of "re-entrant," see the illustrations of polygons, Fig. 3, first year. A plane perpendicular to **V** and inclined at 45 degrees **R** to **H** cuts the axis of the pyramid 1⅝" above the base.

Draw the plan, the fron
an auxiliary showing the tr
develop the frustum. For
lengths of all lateral edges
by the method explained in
setting the altitude wholly w
to avoid confusion of lines.

The problem is identical in every particular of principle with Prob. 1, Pl. 5. The greater number of sides merely makes its solution longer.

Location.—Place the vertex, in the plan, 2½″ R. and 2″ D.; in the front elevation, 3¾″ D.; in the side elevation, 6⅝″ R.; in the development, 7⅝″ D. and 6¼″ R.

Problem 2.—Given, in an oblique position, an irregular quadrilateral prism with oblique ends. Draw, from an auxiliary view, the front, the top, and the side views, and lay out a development.

The greatest diagonal of a right section of the prism is 1⅝″. Axis of prism, 3¼″ long, parallel with **V**, and inclined to **H** at 37½ degrees L. Bases of prism, perpendicular to **V**; lower base inclined 15 degrees R. to H.; upper base, 75 degrees R. to H. Center of axis, C, in front view, 4⅞″ D. and 10⅜″ R. Center of circle for right section (auxiliary view) is on the axis of the front view produced, and 2⅜″ D. Axis in side view, 15″ R.

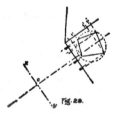

Fig. 28.

Suggestions.—Draw, for the front view, an axis of indefinite length. Thereon locate the center of the circle for the auxiliary view, and draw this view, any irregular quadrilateral resembling that in Fig. 29, one diagonal of which is a diameter of the circle.

From this auxiliary view, the front views can be obtained by a process just that used to obtain an auxiliary view, and the front views are given; see Fig. all points in all views after completing

Development.—The right section always shows the true widths of its hence, the auxiliary view, which, for the may be considered as a right section ma as **x-y**, Fig. 29, gives this necessary info:

In the development, the trace of pla a straight line perpendicular to the la the prism. Draw this trace horizontal On it, set off the widths of the faces and eral edges through these points.

The front view gives the true lengths edges. Using the trace of the plane, line, take the distances from it to the up ends of the edges, and transfer these m corresponding edges in the development respectively above or below the trace plane. By connecting the extremities mined, the development, showing the bases, can be completed.

Next, the true sizes of the bases m This can be done in several ways, by ev auxiliary view, or by construction.

The first two methods, which have explained, are shown in Fig. 30. Note tl are identical. Fig. 31 illustrates the tl i. e., by construction.

In the top view, Fig. 31, assume an with V (as **x-c**) as a base line, and upon vertex of the angles of the polygon in pc

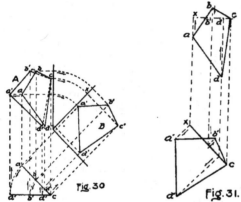

Fig. 30   Fig. 31.

d'. The front view of this base line, coinciding with the front view of the plane, is the true extreme width of the polygon, and also shows the true measures of the widths between the vertices of the angles; Rule 12.

In the top view, the distances a-x, b'-b, and d'-d are taken perpendicularly to V; therefore are parallel with H; hence appear in truth; Rule 12. These give the true extreme depth of the polygon and the true relative depths of the vertices of its angles in reference to the base line.

Now, it is evident that, if the true depths, shown in the top view, be combined with the true widths, shown in the front view, on the base line x-c, by the co-ordinate method (Prob. 5-6, Pl. 2, First Year), the true size of the polygon will be established. Notice that the resulting figure is the same as that found by the first two methods, Fig. 30. Also, by comparison, it will be seen that the off-sets in Fig. 31 (a-x, b'b, d'-d) correspond exactly with the like off-sets from

line x-c' in the auxiliary vi the result in Fig. 31 may be c view obtained by constructio the evolutions being omitted.

In this way, find the true prism. The sides of these sponding lines in the develop ing careful to do this properl; would meet if the developmer

In inking, show all con: the true sizes of the bases in

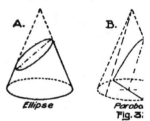

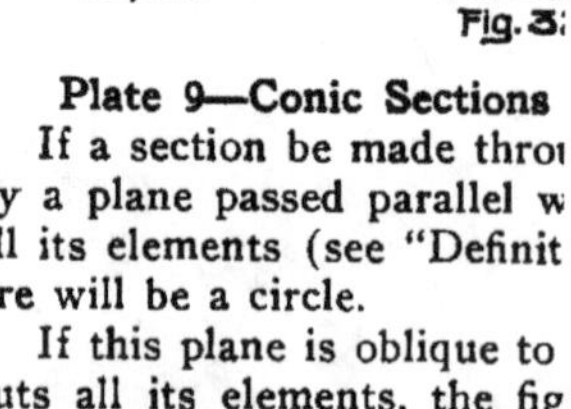

A.   B.

Ellipse   Parabo
Fig. 3:

Plate 9—Conic Sections

If a section be made thro by a plane passed parallel w all its elements (see "Definit ure will be a circle.

If this plane is oblique to cuts all its elements, the fig Fig. 32A.

If the plane is passed pε (cutting the base), the figure 32B.

If the plane is passed par; cone (but not coinciding wit the figure will be an hyperbo

Problems.—Given three, right circular cones, each with a base of 1¼″ radius and an altitude of 3½″. In the top view, the axis of the first is 2⅜″ R.; of the second, 8″ R.; and of the third, 14⅜″ R. The axes of all are 2″ D. In the front view, the bases of all are 7¼″ D.

(1) **Ellipse.**—Pass a plane perpendicular to V and at 45 degrees L. to the base of the cone—i. e., 45 degrees R. to H. Find the sectional surface of the top view, and draw an auxiliary of it showing its true size.

(2) **Parabola.**—Pass a plane parallel with the right element (perpendicular to V) and cutting the base ½″ to the right of the axis. Find the sectional surface in the top view, and draw an auxiliary view showing the true size.

(3) **Hyperbola.**—In the top view pass a plane parallel with V, and with the axis of the cone, and cutting the base ⅝″ in front of its axis. Find the sectional surface in the front view.

Suggestions.—In the top view, divide the base of each cone into sixteen equal parts. Draw elements from these points, in each view. Number these elements. In each problem, the points in which these elements are cut by the sectional plane establish the trace of that plane in the conical surface. The elements of a cone correspond to the lateral edges of a pyramid, the section line passing through points in them being a free-hand curve, however. Section points on elements, the projections of which are perpendicular to HA, will be found by projecting them from the front or the top view to the side view of these elements, as the case demands; but an actual side view is unnecessary in the problems, inasmuch as the right or left element of the front view is the equivalent of a side view.

Developments.—Locate the vertices on a base line ½″ above and parallel w: border line; first vertex, 2½″ R.; seco third, 12½″ R., one element of each devel ciding with the base line.

Locate the traces of the cutting plane true positions of the necessary points t upon the right or left elements of the their distances from the vertices of the c fer these to the elements of the same n developments. Attach the sections, bt bases. Ink all removed portions in dottec

Plate 10.—Practical Problem

For Plate 7 the pupil was provided sioned sketches from which to make w ings of the objects shown.

For this plate he must make his own serting the dimensions, and from these h a complete working drawing of the mod a sectional view, all drawn to a suitabl sketches may be   ee-hand working sket may be oblique views of the objects, rest 18, 19, 21, 25, 26.

A globe or a gate valve, a machinist fitter's vise, a stereopticon, an hydraulic cer's scales, a stock, a Stillson wrench, slide valve and a cylinder, or some othe the physical laboratory or the engine r assigned, at the discretion of the teacher

The drawings should be laid out and e care; all views fully dimensioned; "hatched"; shadow lines and line shading ing; and the plate furnished with a n border line.

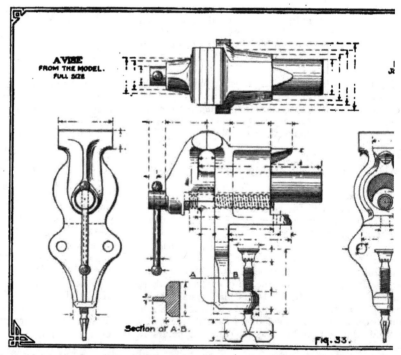

A VISE
FROM THE MODEL.
FULL SIZE

Section at A·B.

Fig. 33.

Figure 33 illustrates how attractive such drawings may be made even when a very commonplace object is used. The pupil should not attempt such complete line shading as here shown, but should try his hand, if at all, only upon some of the simpler parts. Also, Fig. 33 shows how fully, and the manner in which the drawing should be dimensioned.

Or, instead of inking the the pupil may make a traci cloth, as is the more gener Trace on the dull side of th chalk dust.

It is from such tracings t and the original drawings (

THIRD YEAR.
PENETRATIONS.
Introductory Text.

In preparing the text for the subject at hand, Penetrations, it was assumed that the pupil thus far has followed the instructions carefully, and has done his work with understanding. Therefore, further ex planations of general terms, of elementary processes and principles, will be omitted from the succeeding pages. Neither will the particulars of solution be given in such detail as heretofore, nor should the pupil expect them; as the work advances, the pupil must depend more upon his own power to apply principles previously used, thus reasoning out the solutions, and less upon text, in which "it has been thought all out" for him, and which often "shows him how," as well. Needed explanations will not be wanting, however. These are to be followed, step by step, precisely as given.

The workmanship of this year should be decidedly superior to that of the preceding, and in it the pupil should take pride and be ambitious to excel.

Penetrations.—Penetrations, or the cutting or intersecting of one or more forms by another, occur in practice in a large part of the objects of which working drawings are made, as in the penetration of roofs by chimneys, by towers, by dormer windows, etc., in the intersection of pipes and flues in tee, cross, and elbow joints, and in machinery in the joining of hubs and spokes of wheels, etc.

The subject is a direct continuation of that immediately preceding—namely, Sections—inasmuch as the problems in penetrations require for their solution the determining of the various sectional l from the partial cutting of the surfaces by those of another.

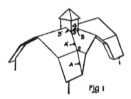

Fig 1

In the accompanying illustration, F tersecting curb roofs crowned by a cu lines **A, A'**, in which the roof surfaces the lines **B, B'**, in which the cupola p roof. To determine such lines of pen necessary to locate points common to t secting surfaces of the objects, in this ca 3, 4, etc. Then these must be joined i gression, as 1 with 2, giving line **A**; 2 line **A'**, etc. Thus a continuous line i its parts following one another in the or cessive intersecting surfaces, as **A**, then **B'**, etc. This sequence is invariable, pro faces or edges of the objects are comme cide to form one.

Two planes intersect always in a hence in the penetration of plane surfac necessary to find only the points in wh of each penetrate the surfaces and edges as shown in Fig. 1.

But in the penetration of one curved with another, or with a polyhedron, it i essary not only to establish the points edges penetrate, but also to locate many other points, intermediate, common to the intersecting surfaces of the objects, because the line of penetration may be made up largely or wholly of curves.

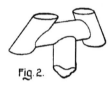

Fig. 2.

Note the curved lines of penetration in the sketch of a chimney top made of cylindrical pipes, Fig. 2, to establish which numerous points common to their intersecting surfaces had to be found.

Statement of the Problem.—Draw the views of the object so as to show their axes in their true relation; that is, assume the forms in a position in which their center lines are parallel with one and the same plane of projection. Then the points necessary to determine their lines of penetration may be found by the use of one or both of two general methods.

Solution—First Method A.—The points may be found from the views on H., V., and S. V., by noting where the edges of each object penetrate the surfaces of the other, and in the incomplete views, by locating and connecting them in correct succession.

Thus, in Fig. 3, the points of penetration of the edges of the horizontal prism appear in the top view, **1, 2, 3**, etc., while those, **A. B**, of one edge of the vertical prism are evident in the side view. Now, by carrying such points from these two views to the front view as shown, and there connecting them, the pupil may easily establish the lines of penetration. This method is particularly applicable to right penetrations of prisms. At this gin Plate 1, referring to the lution as the work advances

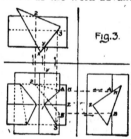

Fig. 3.

B.—The points may be front views with the help o upon a plane perpendicular which presents an oblique the problem.

Although in Fig. 4, the solved, there is sufficient to iliary view is obtained, the the lines of penetration in identical with that just exp seldom necessary to draw th In this case, for example, th prism showing where one penetrates it, is really the else is unnecessary. Henc parts should be omitted, w obtaining them and their pr: This method is particularly etrations of prisms.

Second Method.—The p passing cutting or sectional

secting solids. To insure accuracy, this should be done in such a way as to make the sections in each object simple, easily drawn figures. The points of intersection in the outlines of the figures thus obtained are points in the lines of penetration of the two forms.

The sections may be made in various ways.

A.—By extending the bounding planes of one of the objects, thereby making sections through the other.

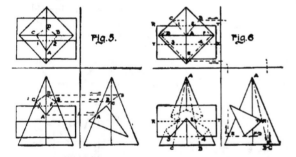

In Fig. 5 we have merely the statement of a problem, showing how such sections are made. Thus, in the side view, the front upper surface of the prism, if produced, will cut the pyramid in the sectional figure **A-B-D-C.** As apparent in the front view, the sides of this figure are actually cut in the points 1 and 2 by the upper edge of the prism, for this edge lies in the sectional plane; hence these are the points of its penetration with the lateral faces of the pyramid. And the lines **A-1** and **A-2,** parts of the figure **A-B-D-C,** are also parts of the line of penetration of the two forms. By continuing this process, the pupil may establish the remaining portions of this line. The com-

plete sectional figures are never needed the following methods of solution; hen should draw only that which is of use, in tion lines **A-B** and **A-C.**

This manner of passing planes is app cially to combinations of prisms and of tl pyramid. In oblique arrangements, an a would be substituted for that on S. V., a done in exactly the same way as above e

B.—By passing cutting planes coincid axes, edges and elements of one or of bo secting forms.

Thus, in Fig. 6, by passing a plane al of **T-R** of the prism, through the vertex mid and cutting the base of the latter, figure **A-B-C** is formed, shown in the f views. The points 1 and 2, in which the figure are cut by the edge of the prism the plane was passed, are the points of p this edge with the lateral faces of the py

This method, with some slight vari application, is adapted to combinations (ery kind. By it, the points of penetrati maining edge of the prism, Fig. 6, may l for the sake of illustration, we will fin different way.

C.—By passing a horizontal plane co the edge **X-Y,** this being parallel with th pyramids, the section made will be a fig similar to that of the base as shown in The points 3 and 4, in which the edge prism cuts the sides of this figure, are penetration. This method is best suited 1 trations of prisms, of cylinders and of ei forms with the pyramid and the cone.

D.—In Fig. 7 is shown a use of vertical cutting planes. In such arrangements, where this method is used, an abbreviated side or auxiliary view supplies all that is needed to establish the figures formed by the sections made. In this problem, in each object these figures are parallelograms. Their intersections fix points in the line of penetration. It will be even

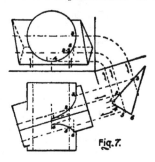

Fig. 7.

more simple to consider the traces of these vertical planes in the lateral surface of the cylinder merely as elements of the cylindrical surfaces which are cut by the faces of the prism, and then to find the points in the line of penetration by the First Method, **A** or **B**, connecting them by curves, of course. This method is adapted to arrangements of forms the elements of the lateral surfaces of which are parallel with their axes, such as right and oblique combinations of prisms, of the prism and cylinder, and of cylinders.

The pupil should so thoroughly familiarize himself with the practical application of these various methods as to have them at his very finger tips, as it were. And he will find that in mastering the subject

of penetrations he will hav
time the essential principles
orthographic projection.

Developments.—The dev
having been laid out entire
How are the lines of penet
therein? To which the answ
ferring the various points in
jections of the objects to
there properly connecting th

Such points as fall in the
give us no trouble to locate,
get always to establish the
lengths of the edges on whic
not already evident, just as v
out developments in Sections

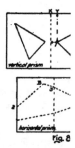

Fig. 8

Prisms and Cylinders.—
the developments of prisms
using the co-ordinate meth
developments of which are
lies in the right face of the
tance **X-Y** (top view) of the
the front edge and the dist

the length taken from the top. In the horizontal
prism the point **A** in the upper face lies in the distance
Z-K (side view) of its width from the front edge, and
the distance **K-A** (front view) of its length taken from
the nearer end. In like manner any intermediate
points may be similarly referenced and transferred to
the developments.

In developing cylinders, utilize uniformly spaced
elements, as lines of reference, and then proceed as
with prisms.

Pyramids and Cones.—Points on conical surfaces,
and those lying in the lateral surfaces of pyramids are
transferred in several ways.

1. Corresponds to Second Method A. By the intersections of the traces of the sectional planes used
in first determining the points.

2. Corresponds to Second Method B. By drawing the traces of the oblique sectional planes, and locating the points on these in exactly the same way in
which points are transferred from lateral edges—i. e.,
by getting their true lengths and establishing the positions of the points thereon.

3. Corresponds to Second Method D. By drawing the traces of the sectional planes when these are
parallel with the base of the pyramid or cone, and
transferring thereto the distances directly from the
view (the top view usually) wherein are shown the
true positions of the points in reference to the lateral
edges or to the elements.

For General Application.—By the intersection
of the traces of the sectional planes and elements of
the surfaces drawn where convenient through various
points in the line of penetration.

The corresponding parts in the lines of penetration in the developments of objects of the same prob-

lem should be equal. This is the proof c
of the work, and the pupil should chec
ments in these particulars before inkin
or the projections from which they wer
is easily done by numbering consecutiv
in the original line of penetration, an
the same numbers to like points in the
see Fig. 8.

It is usual to ink the line of penet:
the development of one object to shov
caused by the passage of the other thrc
the second line is then inked dotte

Definitions—Right and Oblique Pe
the axis of two penetrating forms are
to each other, the penetration is a rigl
If otherwise, it is an oblique penetratio

Regular and Irregular Penetrations
of the penetrating forms intersect, or li
plane, the penetration is regular. If o
an irregular penetration. Thus, penetr:
right regular and oblique regular, or :
and oblique irregular.

The pupil will find that in the st
specifications for various of the proble:
ticulars of size and arrangement are lac
instances, he is to supply these accordi
judgment as to what will be satisfactory
draftsman must know how to proportio:
his drawings.

Lettering.—As exercises in lettering,
design and print each month from Septe
a title for some imaginary drawing, u:
appropriate styles and sizes. See Figs. !
Year, also the accompanying text. Th
these titles will be supplied by the teac

FIG. 9
(with tools)

For May and June, the problem will be the design
of a title page for the portfolio of plates, to be made
upon a sheet of the same size as the plates and to be
bound with them when it is completed.

These designs should be studied out and executed
carefully. They should have a distinctly artistic qual-
ity, for titles are not only useful and necessary upon
drawings, but they also are or should be decorative.
The evidence of nicety of workmanship and of taste
in the design of these particulars as well as in the
making of the drawings themselves, distinguishes the
real draftsman from the mere bungler, the slovenly
one.

In the design of the title page (a book panel), the
groupings of lines and of shapes should parallel the
outlines of the inclosing figure, so that thereby its
rectangular character may be emphasized. And the
shapes used must be related to its outlines in order
that the design may be one distinctly adapted to the
shape of the panel. The lines employed may be simple or elaborated border lines or band-like inventions
corresponding thereto. Or our panel may be subdivided into agreeably proportioned minor panels within some of which figures may be developed. These
band borders and panel figures should be geometric
or conventional rather than organic in character, they
may include appropriate emblems of mechanics, etc ,
and among the shapes to be used the words of the
title also much be included.

The pupil should "spot" his scheme of design—
i. e., study out in sketches the best placing of the various items of the composition before attempting to
develop any one portion.

The lettering may be of any appropriate style.
standard or designed by the student. It should not
be over-ornate. The design may be
various ways, by tinting, by shadow 1
by this means made attractive. See Fi

Text of the Problems.
Plate 1.

Problem 1.—A right square vertical
2½", diagonal of base 2¼", is penetra
square horizontal prism, length 2¾", d
1¼". One diagonal of the base of each
lel with V. H. A. is 3½" D. Axis of v
top view 1¾" R. and 1½" above H. A.
zontal prism is ¼" either in front of
axis of the other. In the front view,
axes are centered 1¾" below H. A. H
intersect 3½" R.

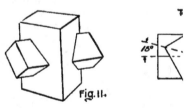

Fig. 11.

Solution.—Use side elevation to find
etration not directly obtainable from
First Method A. Letter both objects
number all points of penetration. See I

Problem 2.—A right square vertical
2½", diagonal of base 2¼", is penetra
square horizontal prism, length 2¾"
base 1½". One diagonal of base of la
with V., and the axis of the prism is par
V. and with H. Penetration is righ

Axes intersect in the top view 8″ R. and 1½″ above H. A. One diagonal of base of vertical prism is inclined at 15 degrees R. or L. to V., Fig. 12. In the front view axes intersect 1¾″ D.

Solution.—Use side view as in previous problem. H. A. and V. A. intersect 9¾″ R. Proceed as before, then lay out developments of both problems in space below and to right of work just done. Locate penetration lines therein, checking them. Read paragraphs on Developments. Ink and tint. In inking the views we quite commonly imagine that the smaller penetrates the larger form, and then omit altogether those portions of the edges of the latter which thus would be removed, as in these two problems we would dot the edges of the horizontal prisms through the vertical, but would omit those of the latter where cut away by the former. Developments are inked correspondingly. Developments are given uniformly a light, flat tint, excepting where openings occur.

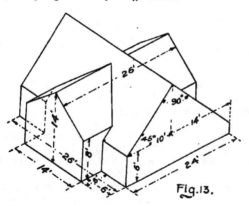

Fig. 13.

Plate 2

Given two irregular, pent bling a barn roof) arranged a sketch, Fig. 13. Longest axis pendicular to V.; that of the both V. and H. H. A. is 5½″

(1) Draw top, front and of axes intersect in top view H. A. In front view base co below H. A. Find lines of per Method A.

(2) Turn top view of (1) prism is at 60 degrees R. to sponding front view, showing

(3) Lay out developments, plate, in space below views, a Show penetration lines, refere

Plate 3

Problem 1.—A right, irre prism and a pyramid. A ri prism 3″ long, edge of base right, regular, vertical penta altitude and 1½″ edge of ba of pyramid is parallel with V prism is perpendicular to S. above the base of the pyramid axis of the latter. Lowest fac degrees R. to base of pyramid view. Base of pyramid 7¼″ cupy left one-half of plate; de Omit bases from development

Solution.—Draw the top, f the pyramid, then the side Produce sides of prism on S.

determining penetration lines on top and front views, and in development of pyramid, Second Method A, using corresponding method in developing. Second Method B or C could be used equally well.

Problem 2.—Practical example. Given a spire in the form of a square pyramid, having dormer windows on two opposite angles; see Fig. 14.

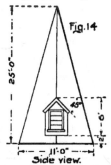

Fig.14

25'-0"

45°

11'-0"
Side view.

Common axis of dormers is parallel with V. and with H. Drawing to occupy upper right half of plate. Base 7¼" D. Scale, ¼" to 12". Upper part of spire broken off in front view to allow space for its top view. Find penetration lines by the method used in the previous problem. Shadow line and dimension; do not tint practical problems. No developments.

Plate 4.

Problem 1.—Oblique penetration of a right, circular cylinder and an irregular prism. A right, vertical, circular cylinder 3½" altitude and 2½" diameter is penetrated by a right, irregular, triangular or quadrilateral prism 4" long, its base inscribed in a circle of 2¼" diameter. Axis of prism is parallel with V,

and inclined 22½ degrees L. to H. front views of cylinder first, then aux right section of prism, as in Prob. 2, F Year, and thereafter draw the other prism. Finally draw developments of omitting bases. Axis of cylinder 3" R.;

Solution.—Second Method D, Fig. 7 cutting planes parallel with V., passing each edge of the prism and two or more two edges. Do not draw the whole cut the abbreviated section or element li traces of these planes in establishing t lines in the developments.

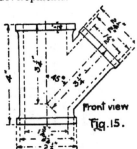

Front view
Fig.15.

Problem 2.—Practical Example. branch, Fig. 15. Draw three views, and line of penetration for both the inner ar ders. Axis of main pipe 10" R., base 7½ draw flanges until penetration lines ha

Solution.—Same as for Problem 1 circumference of the branch into 16 eq pass a vertical cutting plane through e: pipe, drawing, however, only the elem developments.

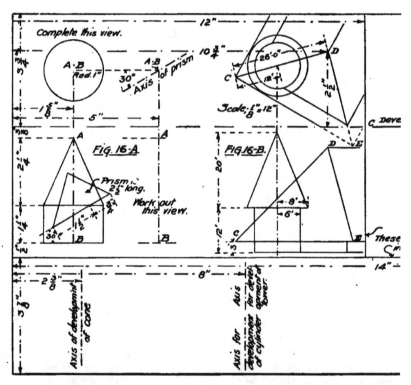

Plate 5.

Problem 1.—Right irregular penetration of a prism with a cone and a cylinder. A cone and a cylinder are arranged as shown in the accompanying location plate, Fig. 16A, and are penetrated by an irregular triangu-

Bk. One

lar prism the axis of which
(1) **Solution.**—Second M
horizontal sectional planes,
lines in the top view. The
cone only.

(2) Transpose the top view just drawn to the position indicated, axis of prism inclined at 30 degrees R. to V., lowest edge to back; get corresponding front view showing lines of penetration thus far found, and complete these lines on the cylinder by the same method of solution used in (1).

(3) Develop the cone and cylinder, but after Prob. 2 has been solved.

Problem 2.—Practical example. This represents the penetration of a hip roof by a cylindrical tower having a conical roof as shown in locat 16B. In the top view, hip **C-D** is at 15 V., and at 90 degrees to hip **D-E**. Als view the hip **C-D** is at an angle of 45 de eave **C-E**, but in laying out the develop: pupil must not assume any of these t angles. This problem is solved in prec way as that used in Prob. 1. Place the where shown, using method of laying o ing with the method of solution used.

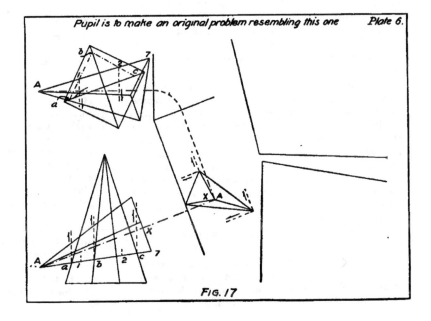

Pupil is to make an original problem resembling this one   *Plate 6.*

FIG. 17

Plate 6.

A right, square pyramid is penetrated by an irregular triangle pyramid; arrangement about as shown in the accompanying illustration plate, Fig. 17. · Draw top, front, partial auxiliary, and developments, working out the lines of penetration complete. Pupil is to determine the exact proportion, arrangement, and placing of work.

Layout.—Draw the top view of the square pyramid, then its front view. Across the front view draw the axis **A-X**, indicating proposed position of the triangular pyramid. Extend this axis, and on a plane perpendicular thereto, draw the auxiliary view of the triangular pyramid only. Then construct its corresponding front and top views.

Solution.—Second Method B. Pass sectional planes perpendicular to H., coinciding with the lateral edges of the vertical pyramid, and like planes perpendicular to V. coinciding with the lateral edges of the oblique pyramid. For example, a plane thus passed through edge **A-7** would cut the vertical pyramid in part in the figure **a-b-c**, top view. The intersections of edge **A-7** with the sides **a-b** and **b-c** of this figure (the traces of the sectional plane) in points 1 and 2 are the points of its penetration with the lateral faces of the upright pyramid. Continuing this process, the pupil may find the points of penetration of all other edges in both forms.

In laying out the developments, it is practicable to use traces of horizontal planes as explained in Developments of Pyramids and Cones, Third Method, or we might apply the Fourth Method equally well.

Supplementary Pro

These· problems, at the d may be substituted for any in the first six plates, but the for another purpose.

If upon the pupil's comp plates, the teacher knows tha intelligently the sectional pl thus far explained, it is reco be given such of the supplem ilar ones in place of Plates ; these processes clear to him.

Problems A, B, and C (Plates 3 and 5. Problems D those of Plate 4, and Problem

Descript

Problem A.—Given the regular hexagonal pyramid pe irregular quadrilateral prism, re-entrant angle. Draw top a penetration lines therein and

Solution.—Second Metho tional planes.

Problem B.—Given the si regular quadrilateral pyrami zontal, irregular quadrilate Draw the top and front view tration, and lay out the devel

Solution.—Second Metho planes coinciding with the fac ject. Exactly the same as P

Problem C.—Given the s square pyramid penetrated r

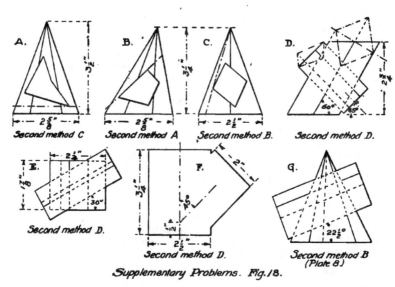

Second method C   Second method A   Second method B.   Second method D.

Second method D.   Second method D.   Second method B (Plate 8)

Supplementary Problems. Fig. 18.

:hombic prism of any convenient size. Obtain the top and front views, the lines of penetration and the developments.

Solution.—Second Method B, by oblique sectional planes coinciding with the edges of the penetrating object.

Problem D.—Given the front view of an oblique, rhombic prism penetrated irregularly and obliquely by an irregular triangular or quadrilateral prism, axes parallel with V. Draw the top and front views, find the lines of penetration, and lay out the developments.

Solution.—Second Method D, by us sectional planes coinciding with the lat each object.

Problem E.—Given the front view irregular pentagonal prism penetrated in obliquely by an irregular triangular or prism, axes parallel with V. Draw the views, get the lines of penetration, and la velopments.

Solution.—Second Method D, by us sectional planes and abbreviated auxilia Fig 7.

Problem F.—Given the front view of a sheet iron, circular cylindrical pipe penetrated irregularly and obliquely by another circular cylindrical pipe, the axis of the second being ¼″ in front of the axis of the first, the axes of both being parallel with V. Get all the views, the lines of penetration and the developments.

Solution.—Second Method D. This is identical with Prob. 2, Pl. 4, but there are no inner lines of penetration to be found.

Problem G.—Given the front view of an irregular, pentagonal pyramid penetrated obliquely and irregularly by an irregular quadrilateral prism, both objects of any size suitable to the plate. Draw the top and an auxiliary view, find the lines of penetration, and lay out the developments.

Solution.—Second Method B, by the use of oblique sectional planes coinciding with the edges of the penetrating object, in combination with a full auxiliary view of the two forms. See Plate 8.

Plate 7.

Given a sphere and an irregular pyramid of any number of sides, arranged about as shown in Fig. 19. Find the lines of penetration in the horizontal and vertical projections and lay out developments of pyramid and of one hemisphere showing lines of penetration in the former only.

Solution.—Second Method B. Pass cutting planes perpendicular to H. through all the lateral edges of the pyramid, and thus find points in which each of such edges penetrates the sphere. For example: In the top view pass a plane coinciding with the edge **x-3**. This plane, as all planes do, cuts the sphere in a circle. Rotate this section and the edge complete, about

the axis of the pyramid unt
V. Then their vertical proj
intersect in two points, a and
of penetration of this edge w
in the same manner to find

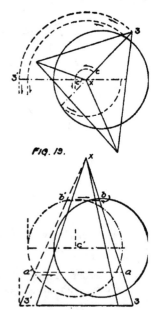

FIG. 19.

edges. To find the necessary
draw several straight lines fro
of each lateral face. These w
ments of those surfaces. Fin
tion by the same method

points of the lateral edges. Should one or more of
the lateral edges fall without the sphere, it may be
advisable to pass several horizontal cutting·planes as
described in Second Method C. In fact, this problem
might be solved by their use, but not so advantage-
ously as by the method already explained.

Development.—After the development of the pyramid has been drawn and the points of penetration of
its edges have been transferred thereto, inasmuch as
all sectional lines are arcs of circles, it remains only
to find the centers for the circles which pass through
these points. This may be done by producing to their
intersection the perpendicular bisectors of any two
chords formed by joining the penetration points on
the same or adjacent edges. To draw the development of the hemisphere, divide its semi-circular profile
into zones of uniform width, four or more, and lay out
each zone as though it were the frustum of a right
circular cone, drawing the sections tangent upon a
common meridian; see Fig. 19a. There is another
method of developing the sphere, but it is even more
of an approximation than this, and of much less practical value.

Plate 8.

A right, circular cone is penetrated obliquely and
irregularly by a right, circular cylinder, as shown in
Fig. 20. Establish the penetration lines in the top and
front views, and lay out the developments showing
these lines therein.

Solution.—After drawing the statement of the
problem, construct an auxiliary view of the combination upon a plane perpendicular to the axis of the cylinder, this view of the latter being then a circle. The
points in which this circle cuts the elements of the

cone are points in the line of penetratio
ing these from the auxiliary to like ele
other views, this line may be located in
is an application of Second Method I

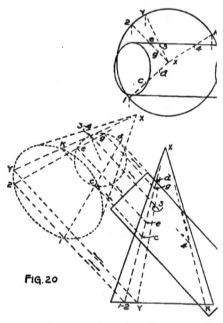

FIG. 20

elements X-1 and X-2, e and g, and c ar
points. A plane passed through the v
cone tangent to the cylinder cutting tl
elements x-y and x-k, will determine th
the points of nearest approach, 3 and 4,

ing parts of the curve of penetration on the front, which we might call, for convenience, the "extreme" or "transition" points. This problem can be made more interesting by cutting the cylinder as shown in Fig. 20a, or it can be simplified by passing the cylinder through above the base of the cone. Also, a prism may be substituted for the cylinder.

Development.—To lay out the development it is best to draw twelve or more uniformly spaced elements on the cylindrical and conical surfaces, and to transfer their points of intersection with the curve of penetration to like elements in the developments as described in the paragraphs on Developments. For the cone see the second method under the heading, Developments of Pyramids and Cones.

Plate 9.

Two, right circular cones, I and II, Fig. 21, of any size, penetrate irregularly in an arrangement similar to that shown. Draw the horizontal and vertical projections complete, and lay out the development of one of the cones, showing penetration lines.

Solution.—To find points in the lines of penetration, pass sectional planes through the vertices of the cones, cutting the cones along their elements. This is an application of the Second Method B. The figures thus obtained will be triangles. The vertex of each triangle thus found will be common with that of the cone of which it is a section. The points of intersection of the sides of two triangles made by one cutting, are points in the line of penetration. All these cutting planes will intersect in a straight line connecting the vertices. The point **B** is the trace of this line in the auxiliary horizontal plane **J-H**, which we have assumed for convenience as coinciding with

the base of Cone I. In orde
needlessly, suppose we pass ;
perpendicular to **V.**, coincid

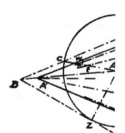

FIG. 21

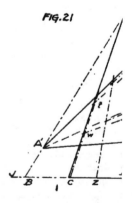

ment of Cone I, cutting C
which we may assume as a
cone. The trace of this pla

is the line **R-N.** The traces of all oblique planes passing through the vertices **A** and **A'** of the cones will be,
in this oblique auxiliary plane, straight lines from **A**
to points in the line **R-N.** And the traces of such oblique planes with the auxiliary **H.** will be lines drawn
from these points on **R-N** to point **B** already established. Keeping in mind that the traces just mentioned lie respectively in the same planes as the bases
of the cones, then the pupil will see that the cutting
of these bases by the traces of the same sectional
plane will determine the triangles the intersection of
whose sides give points in the line of penetration.

For Example.—In the oblique auxiliary plane,
draw any line as **A-X,** top view, cutting the base of
Cone II in points **4** and **6.** Now **A-X** is really the
trace of an oblique sectional plane with this auxiliary.
Its trace with the auxiliary **H.** is **X-B,** which cuts the
base of Cone I in points **C** and **D.** The two triangles
made by this one section are **A-C-D** in Cone I, and
A'-4-5 in Cone II. The intersections of their sides,
elements of the conical surfaces, give us points in

the line of penetration, as **w** and **t,** and (
trace of a plane in the auxiliary **H.** dra
the base of Cone I, as **B-K,** will deter
ment (**Z-A**) in which must lie "extreme"
line of penetration. Corresponding poin
may be found by drawing a trace in the
iary, tangent to figure **1-3-5-7.**

In like manner other points may be f
16 or 20 being needed to establish a
curves of the line of penetration. Incid
be noted that the points **a** and **b,** wher
1-3-5-7 is cut by the right element of
points in the line of penetration.

This method of solution is applicabl
tions of pyramids and cones, and of ot
as right cones. Also it may be used in
of pyramids, the sectional planes being
through their lateral edges; but the met
in Plate 6 is preferable.

Development.—This is laid out in
ployed in the previous problem, Plate 8

Plate 10

ISOMETRIC AND CABINET PROJECTIONS.

We have already noted that in orthographic projection two or more views of an object must be drawn to represent it fully. Also that, in general, in that method of projection, oblique views, being more pictorial, give a clearer idea of the form of the object represented than do simple ones. But, for actual construction, oblique views are not practicable because, properly, they should not be dimensioned, for they cannot be made "scaleable" throughout.

A method whereby we can show correctly in one view the relative positions of the surfaces of an object and also its essential dimensions, would be a great convenience. Fortunately there are two such methods which, within a limited field, may be utilized as substitutes for orthographic projection.

Isometric Projection.—If we place an object, a cube for example, in such an oblique position that its three principal axes are equally inclined from any one of the planes of projection (the vertical is usually taken), we shall find that all will be foreshortened in equal degree; therefore all such axes and, of course, all edges parallel therewith, could be scaled. Now although the ratio of foreshortened to true length is about 9 to 11, in practice all these lines are laid off to scale according to their actual lengths.

Figures so drawn are represented in what is called Isometric Projection—isometric meaning equal measure.

Figue 22 shows the usual manner of making a drawing by this method, viz., two sets of axes extended, one to the right, one to the left, each at 30 degrees to a horizontal, and the third set drawn verti-

cal, all sets drawn to scale. of lines and no angles are sł and sizes.

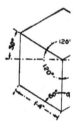

Definition.—Then, Isom method of representation v jections of the three axes of pendicular to each other (. with), meet in angles eithei

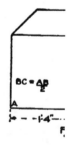

Cabinet Projection.—Thi of the draftsman. It may ł of representation wherein the with those of their surfaces the vertical plane of project

shape, while those of their edges and all dimensions perpendicular to the vertical plane are shown inclined arbitrarily at either 30 degrees, 45 degrees, or 60 degrees, to a horizontal, and usually made but one-half of their actual length. See Fig. 23. Fig. 24 shows an isometric view of the little stool from Plate 6, First Year, while Fig. 25 is a cabinet projection of the same object, both drawn to the same scale.

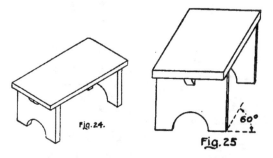

Fig. 24.

Fig. 25.   60°

Notice in the illustrations that something of the effect of perspective is secured, while at the same time the views show the three axes in such a way that the articles properly may be dimensioned.

These methods are used extensively, especially the first, in machine drawing to represent engine parts and details of assembling, and methods of construction and joinery in buildings; also in catalogs and circulars advertising machinery, mechanical appliances, fittings, furniture, etc. Furthermore, they are used quite generally in the making of drawings for the Patent Office.

Small drawings, usually of rectangular forms, are most satisfactory; very large ones are seldom made,

as they are quite awkward to produce a disagreeably distorted appearance.

Very recently a hexagonally ruled pa put upon the market for convenience in metric sketches.

Before working out the problems on t the instruments, the pupil must make, or and submit for criticism, a freehand sk object shown in Fig. 26, or such similar (teacher may direct.

This will not be difficult to do if he mind that the representation is merely ar jection of the object.

Some facility in such free-hand sketcl uable asset to the general draftsman. V ly he is called upon to make such picto from the orthographic projections, or th from the sketches, so that he has need t(miliar with both methods of representati to 26, Second Year, include examples sketches drawn from ordinary projectioi jects.

The free-hand sketches having been a plate may be begun. In this make the cording to the dimensions and arrangem the location chart, Fig. 26. Draw the pro order in which they are lettered. R. P. i ence point", which is the point of beginn ing the outlines of each object. It is lo tances from the upper and the left borde less otherwise stated, make all drawings title of the plate is to be, Problems in I Cabinet Projections, and is to be place rected.

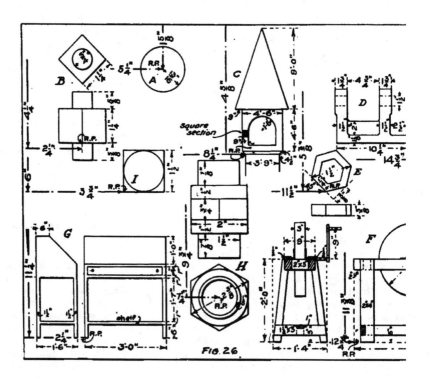

Fig. 26.

The Problems.

Isometric Projection.

Problem 1.—(Fig. A) Make an isometric projection of a horizontal circle, using the four centered approximate ellipse method shown in Fig. 27. Also see Fig. 37, Supplementary Prob. 12, First Year. Fig. 28 is the method of drawing the isometric projection of the circle free-hand.

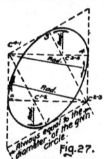
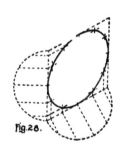

Fig. 27.   Fig. 28.

Problem 2.—(Fig. B) Make an isometric projection of a cube which is penetrated by a cylinder, from the top and front views of the combination.

Problem 3.—(Fig. C) Make an isometric projection of a square belfry tower from its front view. Represent it as if it were above the horizon. Scale, 1/3″ to 12″. Get this from the scale, 1″ to 12″.

Problem 4.—(Fig. D) Make an isometric view of this shaft stand from its front and side elevations. Draw it with its length to the left and its width to the right from R. P. Scale, 3″ to 12″.

Problem 5.—(Fig. E) Make an isometric projection of a hollow, regular hexagonal plinth, one diagon-
al of which is at 45 degrees L. to V., giv front views. Inclose it in a parallelogr Recollect that the length of one side e that of a diagonal.

Problem 6.—(Fig. F) Make an iso tion of a grindstone and stand from its views. Draw it with its length to the width to the left from R. P., which poir lower, left corner of the inclosing rect: shown by dotted lines. Scale, 1½″ to

Cabinet Projection.

Problem 7.—(Fig. G) Make a cabir suitable for a catalog illustration, of a w similar article of furniture, showing the the left from R. P. at 30 degrees acc front and side elevations shown. Scale,

Problem 8.—(Fig. H) Make a cabi of a union pipe joint from its front ai drawn retreating at 30 degrees to the rig and one-half of its actual length.

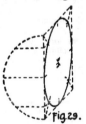

Fig. 29.

Problem 9.—(Fig. I) Make a cabir of a cube having on each face a circle sides. Top and right face to show at 4! to be one-half their actual depth. See I

INDEX

Accuracy, definition of.................................. 4
Alphabets and numerals, examples of.................... 7
Arrow heads, examples of..............................32
Auxiliary or oblique plane of projection, use of.........48
Axes of projection, explanation of....................41-48

Blue prints and tracings................................69
Border lines and corners, design of....................28

Cabinet projection, explanation of.....................87
Cabinet projection, problems...........................90
Checking, developments of penetrations.................74
Checking, developments of sections....................58
Circle, arc, chord, circumference, diameter, radius, tan-
    gent, definition of................................. 6
Circle, to obtain oblique view of.....................53
Circle, to shadow line.................................36
Conic sections; ellipse, hyperbola, parabola, derivation of..67

"D" and "R," meaning of.....................30-49
Data for working drawings............................26
Descriptive geometry, definition of....................41
Developments, explanation of...........................31
Developments of cylinder...............................35
Developments of "penetrations"..........................73
Developments of prisms and cylinders... 73
Developments of pyramids and cones....................74
Developments of regular pyramids......................37
Developments of rhombic prism.........................66
Developments of "sections"..............................58
Developments of sphere.................................84
Developments, practical uses of........................54
Diagram, uses of triangles............................. 3
Dimensions and dimensioning.......................27-32
Dividers and compass, use of.......................... 4
Drawing materials, list of............................. 2

Elements, of cylinder and cone, definition of.............24
Evolutions, of lines, explanation of....................49
Evolutions of solids, explanation of....................48
Extension lines 27

Geometric constructions, layout ⌐
Geometric constructions, **problem**
    to bisect an angle ....
    to bisect an angle, sides not
    to bisect an arc............
    to bisect a line............
    to construct an approximate
    to construct a true ellipse...
    to construct a regular hexago
    to construct a regular octagor
    to construct an oval, axis giv
    to construct a regular pentag
    to constuct regular polygons
    to construct an equilateral t
    to construct a triangle, side:
    to construct a triangle, two
        given ....
    to divide a line into any r
        parts ....
    to draw an arc through three
    to draw a line parallel with
        given point ....
    to draw a tangent to a circ
        from a given point .....
    to draw circles within a giver
        to each other ....
    to draw three circles within
        tangent to it and to eac
    to draw three circles tangent
    to erect a perpendicul r to
        point ....
    to erect a perpendicular at th
    to inscribe an approximate el
    to inscribe a circle within an
        about a given triangle..
    to inscribe a regular pentago
    to inscribe a regular heptago
    to inscribe a regular octagon
    to transfer an angle ....
    to transfer a figure by co-or
    to transfer a polygon by par
    to transfer a polygon by tria
    to trisect a right angle......

Geometric constructions, purpose of.................... 5
Geometric terms, definitions of....................... 5
    angles 5-6
    circles 5
    conical surface and cones.........................24
    cylindrical surface and cylinder..................24
    frustum, definition of............................24
    lines 5
    planes 6
    polygons 6
    polyhedrons 23
    prismatic surface and prisms... 23
    pyramids 23
    solids 23
    sphere 24
    truncate, definition 24

Implements, directions for using.....................2-3-4
Inking, directions for—
    developments of penetrations......................74
    geometric constructions 11
    projection drawings 46
    views and developments of sections 58
    working drawings 26
Introduction, first year............................. 2
Introduction, second year.............................41
Introduction, third year..............................70
Isometric projection, explanation of..................87
Isometric projection, problems........................90

Layout of plates for geometric constructions.......... 8
Layout of plates for projection drawings..............49
Layout of plates for working drawings.................25
Layout of a projection drawing........................44
Leads, pencil and compass............................. 2
Lettering, examples of................................ 7
Lettering, exercises 8-27-29-32-49-74
Lettering, points in working drawings.................31-38
Line shading, explanation of..........................47
Lines, examples of with their uses....................12
Lines, projection or projectors.......................41
Lines, penetration, checking in developments..........74
Lines, sectional, checking in developments............58
Lines, sectional or traces, defined...................56

Materials, drawing, list of........................... 2

Neatness, definition of......................
Numerals and alphabets, examples of........

Oblique or auxiliary planes of projection, use
Oblique views, derivation of..............
Oblique view, from abbreviated auxiliary...
Oblique view of circle, to draw............
Orthographic projection, definition of.......
Orthographic projection, explanation of.....
Orthographic projection—problems, practical
    apparatus, machine parts, etc...........
    base for post...........................
    cap for post, section...................
    check valve, details from section........
    cylinder head, section..................
    cylindrical pipe, developments..........
    engine crank, section ....
    flap valve, details from section..........
    furniture drawings, from sketches......
    oil cup, sections ....
    pump, details from section..............
    skew pins, oblique views................
Orthographic projection—problems, theoretic
    conic sections ....
    cube, section and evolution............
    frustum of cylinder, evolutions........
    irregular pentagonal pyramid, sections..
    oblique irregular quadrilateral prism from
    oblique triangular pyramid, section.....
    rectangular pyramid, simple views.......
    regular hexagonal pyramid, oblique view
    regular hexagonal pyramid, oblique view
    regular pentagonal prism, evolutions....
    right circular cylinder, sections.........

Parallel lines, definition of................
Parallel lines, to construct.................
Parallel lines, to draw with tools..........
Penetrations, arrangement of problems.....
Penetrations, definition and explanation of..
Penetrations, developments, checking.......
Penetrations, developments, inking..........
Penetrations, developments, prisms and cylir
Penetrations, developments, pyramids and c
Penetrations, illustrations of..............

Penetrations—**Problems, practical**—
  method 1-A, barn roofs...........................77
  method 2-A, spire with dormers...................78
  method 2-C, hip roof, cylindrical tower and conical
    spire 79
  method 2-D, "Y" branch; cylindrical pipe...........78
Penetrations—**Problems, theoretic**—
  method 1-A, rt. irreg., square prisms...............76
  method 1-A, rt. reg., square prisms................76
  method 2-B, obl. irreg., prism and pyramid..........83
  method 2-A, rt. irreg., prism and pyramid........77-81
  method 2-B, reg., prism and pyramid................81
  method 2-B, irreg., sphere and pyramid.............83
  method 2-B, obl. irreg., cylinder and cone.........84
  method 2-B, obl. irreg., pyramids..................81
  method 2-B, obl. irreg., cones.....................85
  method 2-C, rt. irreg., prism, cone and cylinder.......79
  method 2-C, rt. irreg., prism and pyramid...........81
  method 2-D, irreg., prisms.........................82
  method 2-D, obl. irreg., cylinders.................83
  method 2-D, obl. irreg., cylinder and prism..........78
Penetrations—**Solution**—
  method 1-A, for rt. penetrations of prisms..........71
  method 1-B, for obl. penetrations of prisms..........71
  method 2-A, for penetrations of prisms, prisms and
    pyramids 72
  method 2-B, for penetrations of all combinations......72
  method 2-C, for rt. penetrations of prisms, cylinders,
    and these with pyramid and cone................72
  method 2-D, for penetrations of forms having parallel
    elements 73
Planes of projection, explanation of 41
Planes, true size of, by auxiliary.....................57
Planes, true size of, by construction... 66-67
Planes, true size of, by revolution.....................66
Plate 1—first year—illustration of.................... 9
Plate 2—first year—illustration of....................14
Plate 6—first year—illustration of....................30
Plate 11—first year—illustration of...................38
Plate 1—second year—illustration of... 50
Plate 4—second year—illustration of..................55
Plate 10—second year—illustration of..................69
Plate 5—third year—illustration of....................79
Plate 6—third year—illustration of....................80
Plate 10—third year—illustration of..................89
Polygons, definitions of............................ 6
Polyhedrons, definitions of..........................23

Principal view, from an auxiliary
Projection, cabinet, definition an
Projection, isometric, definition
Projection, orthographic, definiti

Reading, projection drawings an
Right line shading, explanation o
Rules, orthographic projection..

Scales, use of ....
Sectional or cutting planes, use
Sectional lines or traces, locatio
Sectional planes, uses in penetra
Sectional points, true positions o
Sectional surface, true size of...
Sections, developments of......
Sections, explanation of........
Sections, varieties of...........
Shading, tint ....
Shading, right line.............
Shadow lining ....
Shadow lining a circle..........
Simple views, definition of......
Size and placing of problems, ge
Sketches, for problems........
Sketching, isometric ....
Sketching, shop ....

Tee square and triangles, uses of
Thoroughness, definition of....
Tinting, directions for..........
Titles, design of...............
Titles for plates, second year...
Title page, design of...........
Title pages, illustrations of.....
Traces, definition of...........
Traces, location of.............
Tracings and blue prints ....
Transposed views ....
True lengths of lines, by constru
True lengths of lines, by revolut
True sizes of angles............
True sizes of planes, by auxiliar
True sizes of planes, by construc
True sizes of planes, by revolutic
True sizes of sectional surfaces.

Use of implements, directions for...................2-4

Views, auxiliary or oblique, derivation of..............48
Views, orthographic projections, derivation of........41-42
Views, principal from auxiliary........................66
Views, transposed 48-52
Views, working drawings, explanation of................25

Working drawings, data needed for.....................26
Working drawings, explanation, illustration of..........25
Working drawings, developments, explanation of........31
Working drawings, dimensions and dimensioning......27-32
Working drawings, inking 26
Working drawings, layout of...........................25
Working drawings, numbering points of................31
Working drawings, shadow lining....................12-26
Working drawings, titles for........................27-28
Working drawings, views 25

Working drawings—Problems—
    caster ....
    chimney ....
    cylinder ....
    flight of steps ....
    flower box ....
    magazine rack ....
    pedestal ....
    prism ....
    prism, oblique position ....
    pyramid ....
    pyramid, truncated ....
    shelf ....
    stool ....
    street lamp ....
    wall bracket ....
    waste box ....